HOW THE RHINO
LOST HIS HORN

HOW THE RHINO LOST HIS HORN

CAUTIONARY TALES FROM APPALACHIA TO AFRICA

JACK RATHMELL

First published in 2025 by
Jack Rathmell, in partnership with Whitefox Publishing Ltd

www.wearewhitefox.com

EU GPSR Authorised Representative
LOGOS EUROPE, 9 rue Nicolas Poussin, 17000, LA ROCHELLE, France
E-mail: Contact@logoseurope.eu

ISBN 978-1-917523-26-4
Also available as an eBook
ISBN 978-1-917523-27-1

Jack Rathmell asserts the moral right to be
identified as the author of this work.

While every effort has been made to trace the owners of copyright
material reproduced herein, the author would like to apologise for
any omissions and will be pleased to incorporate missing
acknowledgements in any future editions.

All photographs and illustrations in this book © [CONFIRM],
unless otherwise stated: XXX

Designed and typeset by seagulls.net
Cover design by Gunjan Ahlawat
Project management by Whitefox Publishing

'If you want to build a ship, don't drum up people together to collect wood and don't assign them tasks and work. Rather, teach them to long for the endless immensity of the sea.'

– Antoine de Saint-Exupéry

CONTENTS

ACT I
(THE CAPE CRUSADER)

ACT II
(FEES MUST FALL)

ACT I

(THE CAPE CRUSADER)

SHARKBAIT

I'm sure there are at least a couple of days out of any calendar year where being a disembodied tuna head wouldn't be too bad.

This, however, if you're wondering, was not shaping up to be one of those days. I found myself on a dinghy somewhere off the South African coast, along with a dozen or so nervous, wet-suited tourists, and a handful of less nervous (but definitely not *not* nervous) crewmen. The latter group, desperate to advertise how safe and relaxed we should all feel, wore board shorts and flip-flops in spite of the sterilizingly cold water, and made banal chit-chat to keep us distracted from the fact that our group had paid, in total, several thousand dollars to let ourselves be locked inside a cage and lowered into a feeding frenzy of great white sharks. A feeding frenzy of great white sharks that we, ourselves, had incited, by the way, with the buckets of fermented chum we'd brought along with us.

With the water already churning from the bloodthirsty killing machines wrestling beneath the surface, and more fins still approaching, you'd have thought these chum-buckets would have sufficed, as far as incitement. Evidently not. The captain opened a tackle box and pulled out a coil of thick, weathered rope. Attached to one end was an industrial meat hook, the kind a slaughterhouse might use to hang its slain pigs. He then reached into a vat of grey, cloudy water and pulled out our aforementioned tuna head, which was nearly the size of his own. He held it aloft with a rictus grin, like an orc showing off a scalped battle trophy. He paused for a moment, perhaps hoping for a roar of approval, then lowered the head and smashed the hook down through it with a vicious hammer blow. But all this wasn't just for dramatic effect. This rudimentary fishing lure, we were informed, would actually play a vital role as a focal point for the sharks to converge upon, giving us a better chance of seeing them up close.

'They aren't meant to actually catch it,' the captain explains. 'It's like a matador with the bull. You know, with the red cape?' (Evidently, he had not seen the thousands of internet clips showing bulls ignoring the intended target and deciding to gore their stockinged antagonists instead.)

'Watch this,' he says, and tosses the head four or five metres (about five yards) off the side of the boat. The feeding frenzy intensifies. Then, within seconds, a shark bursts up, breaching almost entirely out of the water, snatching the bait clean off the line, eliciting a gasp from the crowd. The captain sheepishly retrieves the rope, remarking that this was 'highly unusual'. (I notice, however, that he's brought at least six or seven backup heads – perhaps he had seen a few of those bullfighting clips after all.) He quickly turns the charm back on: 'Well, looks like you've picked a good day. They must want to put on a show!' The other passengers whooped and applauded—apparently, we were supposed to be delighted to learn that the sharks were capable of jumping higher than the sides of our boat.

Who knows, though, maybe they knew something I didn't; this wasn't really my element, after all. For those of us who grew up in landlocked places like the farmlands of central Pennsylvania, great whites weren't something we ever had to worry about in daily life. Self-styled 'great' whites, as in white nationalists? Sure. Great white *sharks*? Not so much. When it came to the natural world, our main concerns were things like deer ticks. As far as things that actually ate meat (and therefore potentially *humans*), black bears were as scary as it got, but those were seen more as pests that hung around to dig through garbage; they were – as I'd been reminded hundreds of times – '*way* more scared of us than we were of *them*'.

Funnily enough, I'd never heard the same thing said of great whites. And if such a claim were *remotely* true, there was no doubt we would have heard it that morning, during our welcome presentation. Its intention was, ostensibly, to calm our nerves and prepare us for the expedition. But it had left me with more questions than answers.

ॐ

We'd been told to meet at the small headquarters in Mossel Bay just before dawn, with the plan to drop anchor just after sunrise. 'That's when the sharks are most active,' the receptionist had explained over the phone earlier that week. To my ears, that sounded like the one time of day we definitely *shouldn't* be paying them a visit, but the company wasn't telling me how to do *my* job, so I gave them the same courtesy. Not that I couldn't have used the help: my job, at the time, was as an unqualified, overwhelmed, and increasingly disillusioned gym teacher. I'd been at it for a month or so now, volunteering at an underfunded school in Cape Town through a small non-governmental organization.

The bay was about a four-hour drive from the city, so we'd set off the previous day. I was with two other volunteers in a place like South Africa, doing a road trip alone is out of the question; you've got to work with the roommates you're given. One of them, Chet, was a thirty-year-old frat-turned-burned-out-Wall-Street-bro, who'd picked South Africa for its optimal kite-surfing conditions. His main objectives were, however, was – and I quote – 'to chew bubblegum, and [breed with] as many girls as possible'. (And, as you may have guessed, he was 'all out of bubblegum'.) With us was the far less abrasive Teddy, a congenial, heavy-set Australian, who shared neither of Chet's ulterior motives: he didn't kite surf, and was gay. Although this was only the second day of our road trip, we were already a man down. A *girl* down, actually, as our erstwhile fourth member had been one of Chet's aforementioned women, who had bailed, calling a taxi back to Cape Town after things had soured between them (but more on that later).

We arrived in Mossel Bay just after sunset and settled into our cozy, albeit rather unusual quarters, down on the beach: a retired sleeper-train had been brought to the coast and converted into a hostel, with rooms overlooking the bay. I squinted out across the moonlit water, scanning for any flashes of dorsal fin, before giving in to a restless sleep.

Waking to a damp, grey morning, we walked through the fog over to the dive company's headquarters. Once the other passengers had arrived, the crew dimmed the lights and started their slideshow, explaining how the trip would work and listing the various 'Dos and Don'ts', which included items like: '*Do* keep your limbs inside the cage while you're in the water,' and 'Don't *not* keep your limbs inside the cage while you're in the water.' Despite their suspiciously strenuous warnings, the crew made a concerted effort to illustrate the huge amount of respect and compassion they had for the sharks. These dives, they emphasized, were designed with [human] *re-education*, rather than thrill-seeking, in mind. Their goal was to lessen the stigma these animals suffer from, while raising awareness of all the good they do. 'Sharks aren't predators,' the captain said, 'they're simply *misunderstood.*' We had, apparently, 'absolutely nothing to be scared of'. Indeed, as one of the crewmen gleefully proclaimed, 'Don't forget, you're more likely to die from the Black Plague than be bitten by a shark!' He said this as if we were delivering some sort of unexpected, yet unquestionably heartening news, the way a news anchor might announce that the death toll of some far-flung tsunami had been overestimated by half a percent.

The presentation was then interrupted by an announcement that we were now going to be served a buffet breakfast. A *what!?* Something smelled fishy – or, more specifically, *sharky.*

To my mind, the great whites' plan was staring us in the face: what was about to be wheeled out was our last meal, intended to gain our trust and, just as importantly, stuff us to the gills. In our post-feast stupor, we'd be far more compliant as they rounded us up to be sacrificed. The only thing that *wasn't* clear was how the sharks had managed to get these humans to cooperate. Extortion, maybe? Perhaps organized fish-crime had come into possession of a sex tape featuring members of the dive crew and was now leaning on them to entrap their fellow humans. Who knew; it would be one for the future documentary crew to figure out as they pieced together this gruesome – yet entirely preventable – tragedy. As if on cue, an assistant emerged from a back room carrying a large tray of fresh bacon; several men at the table behind ours cheered.

After breakfast (which I declined: 'Nothing for me, thanks. I'll pick at some of the leftovers if – sorry, *when* – we get back.'), the crew handed out masks and explained the mechanics of the dive. In groups of three or four, we would clamber into the cage (which was, for reference, roughly the size of three medium coffins side by side) as it hung over the side of the boat. The crew would then latch the roof and lower us until we were submerged up to our necks. For insulation, we would be given wetsuits that were almost an inch thick and thus incredibly buoyant; to see the sharks, we would need to push ourselves underwater using the steel bar running across the inside of the enclosure at chest level. This segment of pipe was a repeated point of reference on the 'Dos and Don'ts' list. For example:

'While underwater, *Do* hold on to the bar with both hands at all times,' and its corollary: 'For the love of God, *Don't* use the outer cage itself to push yourself under, otherwise your fingers and hands will be exposed to the [heartbreakingly misunderstood] killers on the outside.' We then learned that each group was going to get about five minutes in the water, although the captain added that we could get out even earlier if we decided we'd seen enough. *Fuck that,* I thought, *I'll be getting my money's worth.* I'd paid top dollar for the right to get flayed alive; this overrode any other concerns.

Lastly, we were given a disclaimer: the sharks approached the boat on over 80 percent of these trips, but sometimes they just wouldn't show up, or would leave after only a minute or two. In such an event, we were assured, the dive company would give out refunds to any unsatisfied customers. Apparently, this policy had started after one group of particularly disgruntled clients had been aggressive toward the crew for failing to summon any great sharks to the side of the little speedboat during their ride. (I noticed that these clients had nevertheless refrained from provoking the crew until back on dry land.)

We stood up to make our way down to the docks. From our vantage point up on the hill, I looked out over the waiting ocean, which the overcast morning had turned an ominous, charcoal grey. Spotting a rocky outcrop a mile or so offshore, with a small flock of birds circling forebodingly above, I pointed out toward it: 'We'd better be careful not to crash. Definitely wouldn't want to be stranded out there.'

'Well, that's actually *exactly* where we're headed. Officially, it's called Seal Island, since they use those rocks to rest when they're not fishing in the waters below. But the sharks wait in the water and hunt them when they jump in. So we like to call it "Shark Island", since they're the ones who are really in charge out there.'

He then chuckled, as if we, who were planning not only to incite a feeding frenzy but offer ourselves to it, could *all* agree those seals were too stupid for their own good. As the crew walked us down to the docks, I asked the captain how the past week's dives had gone. He delivered some more disquieting news: 'Very well. The sharks have shown up every day. But they've almost been a bit too excited. In fact, the main boat suffered a bit of damage on its last trip out to sea.'

He gave no further detail – a silence, which was, in my opinion, deafening. None of us dared to inquire further before he continued: 'Anyway, you'll be in *this* one today.' He nodded over to what I'd presumed was a lifeboat for one of the numerous *actual*, safe-looking boats that surrounded us on the pier.

'Will there be a cage?' I asked, noticing the suspicious absence of one.

'Oh, of course. My assistant Marcus is bringing one of the spares down now. The main one is being repaired. *That* suffered some damage, too.'

The first mate and another crewman soon returned, lugging the cage with them. I couldn't help but notice the duct tape wrapped around several of its frame's load-bearing joints. I recognized the oft-mentioned bar running through the middle, which for some reason looked sturdier than any of the parts exposed to the two-ton sharks. Indeed, the only barrier seemed to be merely some sort of chicken wire – I was pretty confident *I* could have bitten through it if push came to shove. I suggested as much to the captain.

'Don't worry,' he replied, 'they won't actually be trying to get *through* the cage, it's just there for them to bounce off if they come too close.'

'I'll take your word for it.'

'You guys are lucky, by the way – the spare boat actually has a bigger cooler for the snacks!'

I tried to feign some confidence. 'Oh, sweet. What's on the menu?'

A poor choice of words; we all knew what was *really* on the menu that morning. But I'd have paid no attention to his answer regardless, as I was already getting queasy at the prospect of being stranded at sea with a dozen strangers as we sank, if not literally, then certainly into utter delirium from the panic and hypothermia. And so you'll forgive me if the questions flooding my mind were far more pressing than which refreshments were stocked in the cooler. For example, how much panic-induced diarrhoea my wetsuit could absorb before any excess would spray from the neck, wrists, and ankles, further saturating the waters, adding gasoline to a fire that didn't need the help one bit.

❧

What the hell was I thinking? I hadn't come to South Africa to tempt fate. I'd only signed up for this dive – and for this road trip in the first place – because my manhood, my *virility*, had been called into question by Chet, some guy I'd only known for a couple of weeks. Was it really worth risking my life for a near-stranger? Apparently so – and this shark dive wasn't even the dumbest thing I'd do at his behest *that day*.

For a while, I struggled to reconcile it all. This wasn't something Jack would do. I must have been bullied; coerced. This was all Chet's fault, that bastard. And yet, when I look back on that first trip now, having returned to South Africa twice more and spending two years there in total, I'm not sure I would change a thing. So where does that leave us, then, when it comes to the Chets of the world; the self-styled 'disruptors'? With those who push envelopes, throw elbows, and tread on toes, knowing that the havoc they wreak can be the catalyst for good – for growth, reflection, art, and innovation?

My ambivalence here is representative of something a lot of us are wrestling with, I think, as we try to make sense of the world. Inherent contradiction is everywhere; cognitive dissonance is, seemingly, mandatory. With information so available, there's no escaping the fact that everything we are – everything we have and enjoy – came at a cost, be it human or environmental. It's not clear how we should feel about all this. Ashamed? Apathetic? Proud? The Chets, I assume, would say that we ought to be grateful for everything that got us to where we are, lumps and all: sure, the methods for driving society forward aren't pretty, but you've got to crack a few eggs if you want to make an omelette.

Perhaps. But I'm reluctant to let them off the hook so easily. The ends may well come to justify the means in some cases – they certainly would on the day of the shark dive, for instance – but Chet couldn't have known that in advance. My sense is that there's a good deal of survivorship bias: there's a hell of a lot of hardship and strife that *doesn't* get alchemized; a hell of a lot of people for whom things *haven't* turned out all right, and we just don't hear from them because they've been silenced or wiped out. History's written by the victors, and all.

So the questions linger, I think. Which takes precedence: the intentions, or the outcomes? Our external circumstances, or ourselves? By that token, to what degree are we beholden to the past? Answers will, I fear, only become more elusive, as the lines between public and private life, politics and entertainment, art and advertisement, and between past and present grow ever more blurred. As our world has become more interwoven and complex, so too have our stories.

But this is exactly why it's more important than ever to try and take stock; to figure out what got us here, our role in it all. Why we believe, want, and do the things that we do. Not that we'll always like what we find. Indeed, the question of how the fuck, exactly, I, a teenager from Amish country in the US, ended up in Africa in the first place – let alone submerged in rotten seal shit off its coast – is not one with a totally straightforward, nor a particularly flattering, answer.

But we owe it to ourselves, and to the future, to strive for authenticity and clarity in an increasingly oversaturated and confusing world. So it's time for me to come clean. To ... face the music, as it were.

Which means admitting that I didn't go to Africa to 'save the children'. I went there to satisfy a middle-aged German man. I see now that it was never going to work, but I was young, naïve, and ever so star-struck. It's a shame, because I'd been such a huge fan of his work, what with all of his Oscar-winning soundtracks for movies like *The Lion King*, *Gladiator*, and *The Dark Knight*. As for why he chose *me*, I can only speculate; maybe he had a thing for blond, high-school tennis players. The worst part of it all is the fact that he can just go about his life as if nothing happened, as if I don't even exist; he knows it's his word versus mine. But I won't let that keep me quiet any longer.

THE SEEDS
OF DISCONTENT
(OR: FOMO FATIGUE)

I'm referring, of course, to one Hans Zimmer. Our paths had crossed a few years earlier, when he'd sucker-punched me, unprovoked, in front of my family. But wait, it gets worse: the devious little schnitzel-slurper chose Christmas to stage his attack. The *Nativity*, for Christ's sake. You (like my unsuspecting teenage self) might have assumed that a German, of all people, would have been the first one to observe the Yuletide ceasefire. You (like me) would have been wrong.

The year was 2010, and I was a sophomore in high school. With my family watching on, I tore away the wrapping paper to reveal ... my first iPod! A miracle! Until now, my parents (or Santa, rather) had staunchly resisted the early 2000s tide of gadget culture. Even here, I should add, the gift was only offered reluctantly: my dad had won it in a raffle at the office holiday party and had offered it to my mom first, but luckily she'd had no interest. While it's clear now just how prescient their scepticism was, it had been infuriating at the time. For years, I'd had to watch my friends get the newest phones, music players, and video-game consoles, while the most advanced technology I'd been allowed was a knock-off Game Boy from the 99-cent bin of the local Chinese grocery – it had basic versions of *Tetris* and *Snake*, and only lasted a few days before shorting out, to my dismay. But those interminable years of envy and pent-up frustration were already forgotten. I gleefully downloaded all the apps my friends had shown me. One of these, for instance, conjured a virtual lighter to the screen that ignited if you gave it a little shake and we would watch on, mesmerized. (This was a simpler time.)

I then saw that *Inception*, a box-office hit that summer, had released its own official app that would, it claimed mysteriously, 'augment my reality'. I couldn't say no to that. But I quickly discovered, to my mild disappointment, that it offered nothing like the technology used in the film, which had allowed the characters to design and explore complex dreamworlds. Instead, it enhanced the user's world by providing background music, playing different songs based on whatever you were doing at the time – there were about a dozen tracks to unlock in total. Given that Herr Zimmer had written a brilliant score for the movie, however, I convinced myself that this little gauntlet he'd contrived would be a worthy cause and promptly got to work.

The first song required that you hold the device in direct sunlight – thankfully, it was a clear day. For the next, you had to be using a stairway, so I ran back inside and scampered up and down the basement steps a couple of times until the app caught on. The rewards weren't particularly noteworthy, mostly just loops of mysterious, ethereal ambience, but it was an engaging treasure hunt all the same, and I spent the rest of my afternoon scurrying around at *mein neu führer's* behest. Eventually, I got to the last item on the agenda. But when I tapped the screen to load my instructions, the dopamine hit I'd been expecting didn't arrive. All I got was a padlock-shaped notification window:

'To unlock this track, open this application while in Africa.'

I was dumbstruck. *What was this? Some kind of sick joke?* That evening, I forced down some stuffing and a Brussels sprout, but the gut punch I'd received earlier had spoiled my appetite.

'Nobody in our family's ever gone to Africa. I don't even *know* anyone who's been!' I said to my parents. Here I was, stuck in rural Pennsylvania, no trip to Africa in sight. And why would there be? People from Amish country didn't just go to *Africa,* they went to the state farm show to see the thousand-pound butter sculptures (and even these were never particularly memorable, usually depicting some mundane, bucolic scene; a pigtailed little Amish girl petting a life-size dairy cow, for instance). I balked at the injustice of it all; there was no rhyme or reason. They shot a few scenes of *Inception* in California, too, and Paris. Shouldn't I also have to go *there?* I made the mistake of whining in earshot of my grandparents, who needed no excuse to vent their political vexations. They hadn't seen the movie, nor had they any idea what an iPod was, but their eyes lit up upon hearing the suggestion that Africa had received preferential treatment.

'Typical!' said Grandpa. 'Same old story. It's just more of that affirmative action bullshit. By the way, you ought to know your dad got his place in university taken by a black man who had *lower* test scores! We've got to stop pandering to this nonsense.'

'You're goddamn right we do, Grandpa,' I said, slamming my fist on the table. 'Enough is enough. We've gotta take back control.'

I mulled things over deep into the night. While I'd enjoyed bonding with Peepaw (whom I could tell had been beginning to suspect I was 'some sort of queer', given my worrying lack of interest in guns and hunting), I just couldn't shake the feeling that our solution had presented itself all too easily. After hours of tossing and turning, I figured out the real architect of our troubles: Zimmer, the unassuming little music man. I know a weasel when I see one, and I wasn't buying his façade. His compositional talent was indisputable, but this, I realized, was not his real passion. It just paid the bills. What really got his creative juices flowing – what really *got him off* – was the orchestration not of music, but of *social discord*. Indeed, the 'instruments' he most enjoyed puppeteering were *not* violins, flutes, and cellos, but *people*. The more you thought about it, the more it made sense; he was both perfectly disguised and uniquely qualified. It was all adding up: the slimy Frankfurter had set out an ingenious red herring. After inevitably failing his impossible task, we'd blame Africa and its peoples and, in our blind rage, forget who had sanctioned the operation in the first place.

Thankfully, as gadgets of the day only had so much space, I had to delete his Trojan Horse to make room for other apps (designed, I hoped, to do things other than incite racial hatred). After a few months of convalescence, I was able to put the ordeal out of my mind, but the seed had been planted: I didn't know when, and I didn't know how, but I had to prove my parasocial nemesis wrong. I had to simmer the Zimmer.

❧

My glorious stint at the bleeding edge of technology was fleeting; a few months, at most. I spent the following teenage years attempting, in vain, to keep up. Frustratingly, this cycle was more disheartening than never joining the race in the first place; the effects – envy, disappointment, social alienation – certainly felt more acute. I think we were all trained to feel perpetually afraid of missing out, but nobody ever talked about it – not to me, at least.

This struggle was juxtaposed against the lobotomizing backdrop of central Pennsylvania, Amish country. The landscapes are expansive and pastoral: rolling fields of soy, tobacco, corn, manure. The region is dominated by the Appalachian ridge, where ripples of wooded foothills are reserved as state parks and hunting lands. The people are proudly 'blue-collar' and 'hard-nosed', meaning that small-town values like faith, self-flagellating work ethic, and prolific child-rearing are instilled from birth. For many, these norms provide stability and a comforting presence. Unfortunately for me, no such sedating effect kicked in, and my years in farm country left me disaffected and desperate to distance myself.

Departure in any form, however, was explicitly discouraged. Despite the trappings of the modern world, the puritanical scaffolding remained – it's called the Quaker State, after all. Agriculture abounded, but almost all other industry – lumber, mining, and steel production – had dried up after World War II, leaving many of the state's once-thriving towns impoverished and dilapidated. The resulting disillusionment had seeped into the collective consciousness, which led to a suppression of curiosity and intellectualism, as well as a vague distrust of the 'establishment' and the outside world.

Yet it was clear that this rigid pragmatism was no longer rooted in depression-era frugality: the universities that my peers and I were being shunted toward now cost hundreds of thousands of dollars. Moreover, most of the people doing the shunting had all secured their own degrees, jobs, and houses before any of these things required a taking on a lifetime of debt. One of the arguments most often used to dissuade young people from deviating from the school-to-rat-race pipeline was: 'If you leave, you might not have the motivation to come back.' Wilful myopia being a prerequisite for success only struck me as further cause to be sceptical. We were also told that an unemployment gap any longer than a couple of weeks was a one-way ticket to financial ruin. Whether this advice was well-intentioned or not, I felt all this fear and urgency was completely manufactured – these were the oldest negotiation tactics in the seedy second-hand car salesman's playbook.

So, by the time I was a teenager I understood that exploring a far-flung corner of the world was an act of subversion, if not outright rebellion. These bumpkins, Lord love them, didn't know any better. *I*, on the other hand, knew there was more to see and do; that the world was bigger than Pennsylvania. I had a mom from England, which was empirical evidence. This let me reconcile why I was having so much trouble fitting in and with figuring out what I wanted to do with the rest of my life. It wasn't *my* fault, since I wasn't really from here. What also didn't help, I'm sure, was being thrown in the school's 'gifted' programme. This amounted to being told to 'consider yourself lucky', because you had more potential than your peers, but given no tools to actually do anything with this information. You then spend the next decade or so being scolded for failing to figure it out. Oh, well. If I could just break free, everything would fall into place: the skies would clear, and whatever my bright future held would be easy to discern. In the end, I did try university for a year, but I could never get over the lingering notion that three more years of this would saddle me with decades of crippling debt. I knew I had to cut my losses, so, at the first opportunity, I withdrew. It was time to plan my escape.

ò&

My parents said they'd help me where they could, but I'd have to chip in. With only a high-school degree under my belt, however, the only paid work I'd be eligible to do abroad was manual labour – not ideal. If I'd wanted to spend a year picking apples or working in a textile mill, I could have done so at one of the local Amish farms.

For now, this meant I had to follow through with the summer internship I'd signed up for at the beginning of the school year, back when I was still enamoured with the thought of working in politics (aspirations which had lasted approximately three weeks). My role was, ostensibly, to be an aide for my local state representative, but I only ever saw him once. I spent my days working for his chief of staff, an elderly man named Doug. Barring the secretary, however, I was the only other staff member, so his title didn't mean too much. Even so, he liked to pretend we were in some gritty, fast-paced TV drama. Whenever I tried to make conversation, he would wave me away in disgust, claiming that he was 'totally slammed'. *'Hey, rookie, I'm up to my fuckin' neck here. Grab that [phone], would you?'* What he didn't know was that his monitor was reflected in the window behind him, revealing that the only thing he was 'neck deep' in was solitaire. Nevertheless, on his desk he had a framed picture of himself and Hillary Clinton (apparently, he'd helped out on one of Bill's campaigns in the nineties). This gave him a certain cachet, and he knew it. I suffered through ten weeks of mindnumbing small-town politics, helping immigrants fill out census or welfare paperwork, and fielding calls from folks complaining that their request for a disabled parking card had been denied, or that their neighbours' dog was barking too much. It was all perfectly noble work, of course, but it wasn't exactly stimulating, and so I took comfort knowing I'd made the right decision in splitting from the orthodox political-science track. Bigger and better things were in store for yours truly (e.g., brokering world peace, saving the rainforests, etc.).

But it soon became clear that there weren't too many opportunities for a totally unqualified teenager to reshape the global political landscape. When I looked into internships or entry-level roles at places aligned with my 'limitless potential' (read: delusions of grandeur), I found that they were only interested in candidates with not just perfect grades and résumés, but with years of experience in the field. How was it possible to get experience in the first place if every role required you to already have experience? More troubling than this fun little paradox, though, was that I couldn't have gone even if I'd been accepted, because none of the positions at blue-chip places like the UN, or IMF, or wherever were paid, nor did they even cover your travel expenses. You had to move to Brussels, or Geneva, or some other cosmopolitan hub, and find your own room and board for the several months you were volunteering. All in, you were easily looking at tens of thousands in expenses, just for the privilege of putting 'UN' on your precious résumé. Didn't this mean that the people in charge of global affairs were almost exclusively the sons and daughters of people who could afford to finance their children's pursuits? The candidate pool would hardly be representative of anything but the world's richest few percent. Not to worry, the site said, evidently having read my mind: every year, they gave out a few paid internships to a handful of the developing world's best and brightest, allowing them to make the once-in-a-lifetime journey to the Big Apple. (I shuddered to think what Peepaw would have made of that.)

My dreams of saving the world were circling the drain. It wasn't clear how any young person without rich parents (or straight A's since birth) was supposed to do anything remotely interesting or meaningful with their lives. I'd missed the résumé-fluffing boat, it seemed, and didn't want to have to waste the next decade or so as an unpaid chaiwala just to build up 'experience'. Surely, there had to be more to life; one corner of the earth where you didn't have to blow tens of thousands a year just to keep pace with your peers, where it didn't matter who liked or commented on whose social-media post, or how long someone took to text back. Somewhere where 'a comfortable life' and 'meaningful work' wasn't an either/or proposition. I didn't have a great idea why all of us were under all this pressure, and until I had some more answers, I knew I'd have to go off-piste. The way I saw it, we needed to cut through this performative, red-team/blue-team gum-flapping and mud-slinging and do some real good for those that needed it most. We needed boots on the ground. And that's how I ended up in the Peace Corps.

My version of the Peace Corps, that is. I looked the real one up, and got as far (two and a half paragraphs, to be exact) as 'two-year deployment' before I closed the page. *And only one day off per month?* It took itself a bit seriously for my liking. I liked the idea, in principle, but I'd need to start with something a little less daunting. That's when I learned about *voluntourism*: you got to pick where you went, what kind of work you wanted to do, *and* how long you stayed. Half exploring the world, half saving it: the perfect plan. I trawled through dozens of websites that advertised various roles depending on the needs of a given region or school: teaching English in Vietnam, animal rehabilitation in Tanzania, computer education in Guatemala, childcare in Kathmandu. I was giddy with excitement; the Third World was my oyster! This must have been how Mother Teresa or Princess Diana had felt. You want something that looks good on a résumé? How about 'single-handedly saved the world's children'? How 'bout *them* apples?

WANDERLUST™

Until my research began, I'd figured volunteering work would cost nothing but my
time – the Peace Corps had even promised a stipend – but it soon became
clear that an outlay would be unavoidable. All of these programmes required
you to organize your own travel and
cover your day-to-day expenses, so
you'd be paying for a basic lunch and
dinner, as well as a bed in something
like a youth hostel. Fair enough; but I then began to notice a
peculiar hierarchy in these directories of 'benevolence': the
roles demanding the most money didn't seem to be the ones
providing the most aid. As these organizations were all based
in countries where the cost of living was under five dollars
a day, it was therefore confusing to see that rescuing baby
sea turtles in Bali or Costa Rica would cost many thousands
of dollars a week, while that same fee would be enough to
cover working in a childcare centre on the outskirts of, say,
New Delhi for six months. I'm sure accounting justification
for this could be produced, but I suspected it was no coincidence: these companies knew that baby turtles presented

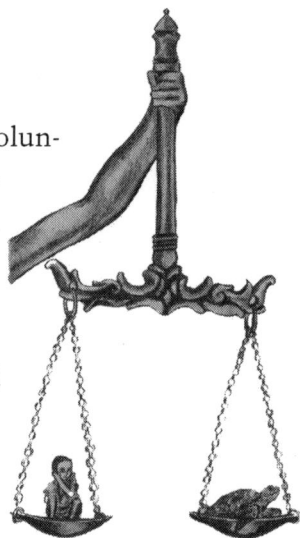

a unique opportunity for burnishing one's personal brand. Feeding some Indian orphans, while still scoring you a *couple* of points, didn't offer quite the same glamour factor.

The organization I ended up choosing had been mentioned to me by a cousin who'd gone abroad a few years earlier. It seemed legitimate enough, with a headquarters in New Zealand that oversaw dozens of smaller charities scattered throughout impoverished countries around the globe. I filtered the list of vacancies by 'sport', as this was the only field in which I had any teaching experience, having coached tennis since high school. The only jobs available were in a seaside suburb outside Cape Town, South Africa, called Muizenberg. It was known for its good surfing, and for its rows of quaint, brightly coloured little beach huts, which reminded me of the ones that lined the shore near my mom's home town in England – I took this as a good omen. I could pick between being a gym teacher at a local school or a surf instructor. As appealing as the latter sounded, I struck it from contention. I'd never surfed before and didn't trust myself to learn on the job. (I have at best, a limited relationship with my centre of gravity: I didn't learn how to use a playground swing until I was well into my twenties, but that stays between us.)

I still didn't know too much about South Africa itself. We had a family friend who'd gone during the 2010 FIFA World Cup and had raved about his experience. And I liked its music – or at least my idea of it: growing up, my parents always had Paul Simon's *Graceland* album playing on our home stereo.

More abstractly, I was drawn to the notion that this would be a bona fide adventure: it was almost ten thousand miles away from home, and, as a bonus, in a whole other hemisphere. If I were to visit, of course, all this exoticism would therefore be conferred upon me, which would do quite nicely indeed. I'd been longing for this fresh start. My year at university had offered that to some extent, but the campus had only been an hour away from home, plus I'd shared with a childhood friend. I'd still very much been my 'central Pennsylvania' self, as had my peers, and I wanted to leave all that behind. I figured this sort of trip would self-select and anyone interested would be operating on a similar wavelength. Ego, neuroticism, status anxiety, and all our other bad habits would *surely* need to be checked at the door, leaving room only for things like compassion and positivity. Last – but certainly not least – this was my chance to finally prove to Hans Zimmer that I wasn't a white supremacist. So, with my arbitrary criteria having been met, my Rumspringa could commence.

Still, I remained wary. I was well aware that this genre of nonconformity had been done before – exhaustively. Any reference to gap-year travel carried with it a certain cliché: trust-fund kids with dreams of 'seeing life from a new perspective' and 'making a difference' during their month or so in Kenya or Fiji. Or the unshaven backpacker, replete with gauged earlobes and a wrist full of tatty music-festival bracelets, who was fed up with 'taking it from The Man'. Or the guy who spent two weeks in Central America and now earnestly affects a Spanish accent when pronouncing any *remotely* relevant country or word (e.g., rolling the 'r' in Nicaragua, or even just when ordering at Taco Bell).

As I researched, though, it began to feel like this triteness was unavoidable. Concepts such as *Wanderlust* and *The Road Less Travelled* had been commodified as personality traits and personal branding aesthetics. People were obliged to prove they were a world traveller by *performing* as one, by posting photos, blogs, and videos every moment of their trip. The need for documentation and validation was (and still is) *especially* urgent when it came to virtuous deeds. As the proverbial tree that falls in the forest without anyone there to hear it, if you'd fed some orphans the other day but forgot to film yourself ladling their porridge, did your act of charity even happen?

But these were an ancillary issues. The overarching questions remained around what this type of trip ought to *be*, what it should *mean*. I was, after all, an imposter, a self-interested fraud; I wasn't someone who'd ever go to Africa to 'rediscover their roots' or 'save the children'. How could I, in good faith, hope to take up my burden, as it were, given all this crippling cynicism?

Valid concerns, but in the end they were no match for the siren song of Mama Africa. I had sights to see and a self to actualize, baby. *So get that exotic passport stamp cocked and ready, I'm comin' in hot.*

❧

I spent the summer monitoring flights to South Africa from every airport on the east coast of the US, tracking and cross-referencing their price fluctuations across multiple websites. After almost a month of deliberation, I chose the cheapest option, which saved me ... forty dollars. Nevertheless, it gave me a feeling of profound satisfaction to know I'd outwitted the behemoth of the airline industry; I'd never been able to capitalize on a market inefficiency before.

Sadly, my reign as apex predator of the free market was short-lived. My hubris had led me into a trap, adding time, expense, and discomfort. I soon learned that departing from New York City was impractical and inconvenient rather than swanky and cosmopolitan. Between all the train tickets and other expenses involved with getting to the terminal, every penny I'd so cleverly saved in my earlier dealings was now gone, (along, to my dismay, with 150 more of my precious dollars). And the flight wasn't even direct to Cape Town; there would be a short layover in London.

On the second leg, an overnight flight, I sat next to a black South African man, and we soon struck up a conversation. He'd been visiting his daughter, who now lived in the United States and had paid for his trip; his journey there had been the first time he'd ever flown. I peppered him with questions about our destination (his home city), before the conversation shifted to politics. For over thirty years he had been loyal to Nelson Mandela's party, the African National Congress (ANC) which had been in power since South Africa's democratisation in

1994. However, the man told me, many who'd initially been hopeful that the self-proclaimed Freedom Party would lead the country into a unified, prosperous future had become jaded.

'The biggest problem today is corruption,' he said. 'It goes right to the top. They're very clever about it. They give big contracts to all of their friends, and they all keep getting richer.'

'What kinds of contracts?'

'Everything. Construction, mining, electricity.'

'Have you ever thought about voting for a different party?'

'No. We can't. The ANC isn't perfect, but there's no other option.'

'What do you mean? There's only one party?'

'No, we just don't trust the others.'

'I thought the ANC were the good guys – what happened?'

He gave me a 'there aren't enough hours in the day for *that* story' chuckle.

'But what about Mandela? He was a hero.' I asked.

'Yes, but he was only one man, and he was old. All of his friends let the power get to their heads. Even his wife. Everyone wants more, more, more.'

'What started all this, do you think?'

He gave me a brief crash course: Mandela had managed to oversee the country's democratic transition with remarkable poise, advocating for peaceful cooperation and forgiveness rather than resentment and hostility. Thabo Mbeki had the impossible job of succeeding him in 1999, but did a decent job keeping things chugging along – barring his being an AIDS denialist, which is said to have cost hundreds of thousands of lives. (Oh, and it was also under his leadership that the country's power supply began to fail. *Other than that, though* ...)

As long as Mandela, who remained the smiling face of the country even after his presidency, was still kicking around, any background concerns could be minimized. But after he died in 2013, the sobering reality could no longer be avoided. It became clear that South Africa's institutions had been overrun by a horde of inept and/or self-serving actors, creating a vicious cycle as these actors further ignored, or actively crippled, regulation and oversight.

But the rot had set in even before the country's democratization. When the Guptas, a family of Indian plutocrats, realising that the fall of apartheid was imminent, they travelled to South Africa, having spotted what promised to be a rich vein for profiteering. Regime change, they understood, would necessitate new institutions and systems which they could shape to their benefit. Their man on the inside was Jacob Zuma, an ANC careerist who, by the sounds of it, was more concerned with accumulating power than wealth (not that the latter wasn't a priority, too). He was more than happy to let the Guptas take the lion's share of any profits as long as they helped him climb the political ladder – which they did: he went on to win the country's presidential election in 2009, taking over from Mbeki. Even more impressive was that this election came just three years after he'd been on trial for rape.

He'd beaten the charges, claiming the affair was consensual. For good measure, he'd also reassured the court not to worry, as he'd taken a shower afterwards to lower his risk of catching HIV (which he and the woman both admitted to knowing that she was carrying beforehand).

After dinner, I drifted off, to be awoken by the captain's voice over the intercom: we'd be landing in about an hour. I opened my window blind to see what kind of terrain we were dealing with. *Would we have to stay in a holding pattern while we waited for the herds of zebras and gazelles to clear off the dirt runway? Would we be kept on the taxiway while machete-wielding local warlords negotiated with the crew?*

When I looked down, though, I didn't see an endless expanse of dunes dotted with oases encircled by palm trees. No Bedouin tents or trains of camels, no black women carrying baskets of fruit on their heads. The only aerial footage of Africa I'd seen were those clips in nature documentaries, tracking shots from a low-flying helicopter of a herd of water buffalo being pack-hunted by a pride of lions across the savannah, that sort of thing. There *were* low, rugged mountains but, to my confusion, they were … snow-capped? *What kind of desert was this?* Had we changed course overnight to head eastward to instead fly over the remote foothills of Pakistan or Nepal?

As we got closer to Cape Town, the snow disappeared, but the mountains continued, and the coastline came into view. Swatches of gleaming white did remain, although these were now where the ground was flat, making up of miles of solar-panel fields, like you'd see out in the Mojave Desert. It was nice to see that this country had invested in sustainable energy. I asked my neighbour about them.

'That?' he said. 'That's a township.'

'Oh. Like a neighbourhood, you mean?'

Back home, a township was what we called a given municipal zone or school district; there was no socio-economic stigma. Here, though, he explained, the word carried a vastly different meaning. Townships were slums; tens or hundreds of thousands of people crammed into crude, makeshift huts, sheds, and shelters built from whatever materials they could find. Sure enough, as we descended, I realized that the 'solar panels' were actually just sheets of corrugated tin, some of which happened to be newer and therefore reflective. These shiny panels were few and far between, however; most were rusted, filthy, or not even metal, just sheets of tarpaulin or plywood.

'Who lives there?' I had a feeling, but wanted to make sure.

'Blacks.'

Over the decades, huge swaths of the black population were forcibly removed from the cities and relocated to designated areas of land nearby. Though installed during apartheid, millions still inhabit these slums, and the borders failed to expand over time in proportion to the population inside.

'And some coloureds, too,' he continued, 'but they mostly live over there.' He pointed to a low sprawl beyond the township.

Coloureds? This threw me. He'd repeated himself, hadn't he? That word, too, I'd heard back home, though it had been used, by certain people, interchangeably with 'black' (along with a couple other epithets, too). I decided to ask him to clarify before I risked getting too comfortable. *Coloured*, it turned out, was a distinct ethnic group more or less unique to South Africa. When the Dutch settled on the Cape in the 1600s, they brought with them slaves from other parts of Africa as well as some from Malaysia. These groups mixed with the local tribes, and over time a distinct population formed. They weren't, he explained, simply mixed-race. Because of their complex ancestry, they had their own, recognizable phenotypes. Of course, nothing was concrete: some looked slightly more Asiatic, with straight, glossy black hair, some more European. They even had their own accent and spoke in their own patois, a combination of English, Afrikaans, and Malaysian, but with some bits and pieces from the regional African languages thrown in, too. The explanation for their not living in townships was simple: during apartheid, they had ranked slightly higher than 'black' in terms of rights and social status. The divisions between where they lived were not just physical; they were baked in at a social level, too.

The man went on to tell me he had lived in or near Cape Town almost his whole life, but he'd never visited Muizenberg, which was only a half-hour drive from the heart of the city.

'I've seen pictures, but it's a town mostly for surfers and the English, not people like me. But it would be nice to go one day. I love the beach. But I don't know how to drive, and taxis are expensive.' Shortly after arriving, I learned a round-trip

35

fare from where he lived would cost about two dollars; I was glad I didn't mention my snafu with the plane ticket.

I heard the landing gear come down and felt the plane's nose drop. *Fasten your seat belts, stow your tray tables, and put your seats in the upright position for landing.* The upright position for *what*, now? This couldn't be right. Usually, at this point in a flight, you'd look outside to see more than enough space. A barbed-wire-fence-ringed airport in the middle of nowhere; some suburb of whatever city you were heading to. Here, however, our final approach was taking place not over manicured grass or a criss-cross of service roads or taxiways, but over one of the very townships I'd just been learning about. *What the hell had I gotten myself into?*

PHI KAPPA ZULU

Thankfully, a normal-looking runway appeared at the last second, but the tone had been set: the boundaries between what was familiar and what *wasn't* were going to be different than back home. The ride to Muizenberg would crystallize this further.

New volunteers were asked to arrive on either the first or third Friday of the month so that the management team could better coordinate our transfers from the airport, meals, groceries, and sleeping arrangements. I was the only one arriving from London, but there were a few others in my intake group. We loaded up in a taxi van and set off. The others fell asleep almost immediately, giving me the chance to take things in. As soon as we leave the airport complex, we find ourselves driving through, or at least alongside, the township. There is no route to the city that avoids this. I wonder if it serves as a sort of tax: you cannot come and go – you with your highfalutin lifestyle – without acknowledging the elephant in the room. After a few minutes we reach the coastal road that traces the northern side of False Bay, the body of water created by the Cape as it juts westward before forging south for a few miles toward Antarctica. The road thus provides an irresistible panorama as it parallels miles of uninterrupted beach, with the mountain ridges of the mainland and the Cape bookending the cobalt of the ocean to the east and west. A breeze rearranges the bay's surface; oblong and crescent-shaped swatches drift to and fro, migrating for a few seconds before losing momentum and fading away to be replaced by another nearby.

On the other side of the road, the land is covered for miles by the township. Compared to the sequined ripples on the water nearby, the tin roofs are far more static targets for the sun; they can do nothing but absorb its rays, which beat down indiscriminately. A hazy mirage hangs a second, distorted horizon above the patchwork sprawl, interrupted only by the telephone poles planted at random intervals; these are the township's tallest structures by some height. Between them run bundles of thick power lines. From their transformers, handfuls of wires hang down, maypole-like, to the surrounding houses. The driver confirms that this arrangement is as unreliable as it looks. During the winter, South Africa's rainy season, the haphazardly configured poles and lines are easily damaged. Given the lack of reliable drainage and sewage, flooding is common, making repairs complicated and dangerous, leaving infrastructure and homes ruinously neglected for months or years.

The road continues. A group of boys play soccer in a dirt clearing, deftly avoiding a cow grazing in the middle of the lot. A stray dog follows two men traipsing down the shoulder of the highway toward the city. On the Atlantic side, the dense shrubbery leading right down to the water gives way to an immaculate beach – which is strangely deserted. When I ask the driver why this is, he chuckles.

'Oh, it's far too dangerous to stop here.'

But there didn't seem to be a bad guy in sight, I contend.

'Yes, well, they hide in the bushes. They'll steal the car no problem.'

The empty road, narrowed to one lane by the encroaching sand being blown inland from the shore, and the neck-high, impenetrably dense shrubbery running along either side of us, now seem eerie rather than tranquil. I can't help but stare intently at the passing greenery, trying to pick out the carjackers waiting to strike. Just a few hundred metres farther down the road, a gated housing development of large, suburban-style model homes is under construction, set to be finished in just a few months. The roadside billboards advertise space, luxury, and panoramic sea views, all just a short drive from the city's central business district.

Minutes later, we were dropped off outside the organization's headquarters, a converted bungalow that housed about a dozen volunteers and had a small office from which the administration team worked during the day. As we unloaded our luggage, I could barely contain my nervous excitement. My adventure was well and truly underway. I'd made it all the way to the farthest tip of this massive, mysterious continent, ready to explore the land while forming lasting connections with other young people from exotic corners of the world.

The outer gate buzzed open, and we filed toward the house. *No turning back now.*

The volunteers welcomed us inside, and I took stock of the group (of thirty or so) as we began to mingle. I had steeled myself for nothing, it turned out. Nearly everyone was white, from North America or Western Europe, and in their late teens or early twenties. At least two-thirds of the group were women. I don't know what I'd been expecting, but it mostly just felt like I'd crashed a homecoming party at, say, Delta Pi.

Sure enough, the first guy I talked to happened to be from a town next to mine in Pennsylvania. Just my luck: I'd spent two thousand dollars and flown ten thousand miles just to reenact a scene I'd been part of at every birthday party, barbeque, and social gathering for as long as I could remember. Here I sat on plastic patio furniture, drinking shitty beer and lamenting the horse-and-buggies we so often got stuck behind while driving through Amish Country. When I went inside to refill my drink, I passed by the living room and noticed a small, sombre group of girls. They were still staring at the floor in silence on my way back through. I popped my head in.

'Hey, guys. Everything OK?'

'Yeah, it's just that … it's Brittany's last night,' one of them said, without looking up. 'We won't be able to go on without her. What are we going to do?'

'Oh, that sucks. Sorry, but who's Brittany?'

Brittany had been their co-volunteer for the past three weeks and was due to fly home the next day. I was invited to her Last Supper at a curry house nearby, a tearful affair during which she received gifts from her more devout acolytes and wondered aloud with them whether life would be worth living upon her return home. Testimonials celebrating her bleeding heart were tearfully exchanged, one example being that she'd let another of the room-mates keep one of her jackets. (I dug a little deeper later on. It turned out that Brittany had wanted to buy a jacket in South Africa but hadn't found one she liked – 'Vera (Wang, allegedly a close family friend) would straight-up puke if she saw some of the shit for sale down here!' – so had spared no expense, having her parents ship one of her sable fur coats overnight from their family home in New York City. One couldn't put a price on looking good for the orphans, it seemed.)

At any rate, I paid my respects, finding myself genuinely disappointed that my path had so narrowly missed that of such an inspiring figure. We walked home, and I waited a few hours for the shiva-sitters to progress through their stages of grief before I broached the topic again.

'So, what made your relationship with Brittany special?'

'It's impossible to describe. We were … Soul Sisters.'

I waited, desperately searching her face for any sign she had delivered this line with tongue-in-cheek. No such luck.

Midway through the night, I found myself being aggressively flirted with by a drunk blonde from Connecticut. I tried to feel flattered but, as I'd already seen her make her way around each of the other new guys, I knew this was little more than a last-gasp attempt. I heard her out all the same. Having barely introduced herself, she began laying the groundwork:

'I just had to get out of the States, and I really wanted to go somewhere far away.'

'Yeah, me too!' *Some common ground, perhaps?*

'Yep, I couldn't last one more day. I needed to start a new chapter.'

'Amen. What made you feel that way? Fed up with school? Work?'

'Well, no. It was my boyfriend – sorry, my *ex*-boyfriend. He was a psycho. We fought all the time.'

With this revelation, I realized that ours would not be an exchange of ideas about the joys of expanding one's geographical or spiritual horizons. This was going to be a 'trauma dump'. I offered my sympathies, hoping to de-escalate the situation: 'Man, fighting can be so draining. My ex-girlfriend and I used to argue a lot, too. Well, at least you're here now.'

'Oh, ha! No, no, I meant we used to *fight* all the time. Like *fight* fight. He used to beat the shit out of me! The last time was in the school cafeteria. He got expelled for that one. But I held my own and got a couple of good punches in. It felt amazing.'

I found myself feeling terribly concerned: for her, of course, given her unenviable situation, but mainly for myself, considering the corner I was now backed into. My face must have betrayed this, and she tried to course correct.

'Don't worry, I have a restraining order against him now.

But anyway, I told myself he'd be the last black guy I ever dated. I need a fresh start. I'm fed up with chocolate, know what I mean? It's time for some mayo.'

She now tried adopting a seductive smile, to disastrous effect. My best guess was that she'd seen somewhere that biting one's lip was sexy but had stopped reading before finding out which lip. As a result, she looked to be doing an impression of an English bulldog's underbite. The decision to travel to Africa to avoid black guys wasn't even the most perplexing part. You need a sexy-sounding foil to chocolate, and decide to go with *mayo,* a condiment made of vinegar and egg whites, when *vanilla* was right there? Even whipped cream would have worked.

The night wore on, and people filed off to bed. I downed what was left of my drink and prepared to follow their lead. It hadn't been a particularly promising start on the whole 'adventure' front, but there was always tomorrow. But then, for a minute, it looked like we were going to get some real excitement after all. I heard murmurings between a few of the tenured volunteers, one of whom ventured into the house to check on a guy named Matt. I asked one of the Soul Sisters who that was.

'Poor guy. His friend Michel got stabbed a couple of nights ago.'

'Sorry?'

'I know, it's confusing. It's French. It's their version of Michael, apparently.'

'No, I know. Did you say he was *stabbed*?'

'Oh, yeah. But it's not a big deal. He's alive and everything, he just had to fly home.'

Matt emerged and came over to greet us. One of the girls observed that he was wearing the victim's hoodie.

'*Fuck yeah*,' he said. 'Check it out, it still has the knife hole!' He twisted to his left to show us a jagged, mortal-wound-sized hole in the kidney area.

The volunteers then explained to us newcomers that the main house had been broken into earlier in the week by a couple of local burglars who'd learned that the occupants were young, naïve travellers who'd brought all sorts of expensive gadgets. Michel had unwittingly stumbled out of his room to use the toilet and came face-to-face with two ski-masked intruders holding laptops that had been left out in the common room. Still groggy, he'd blocked their way and tried to question them, at which point one of them lashed out and caught him with the knife. The men then bolted, escaping out the back door and over the fence. This was all recounted with a puzzling nonchalance.

'There are different rules here, that's all,' said one of the Sisters. 'That's just what happens. It was our fault for forgetting to turn on the alarm before we went to bed. One girl flew home last week after getting her laptop and passport stolen from her at gunpoint. But come on, she walked to the internet café along the same route every day. What did she think was going to happen?'

To us newcomers, the girl who'd been mugged didn't *sound* particularly stupid; she sounded like the helpless victim of a terrifying gunpoint robbery. We looked at each other in disbelief, not least because *none* of these incidents had been disclosed by the administration team.

❧

Work was to start on Monday, but I had no clue what to expect. Getting to coach some tennis would have been ideal, but I had a feeling I wouldn't be doing much of that here: my Google Earth reconnaissance of the school's surrounds had shown little more than a barbed-wire-lined dirt patch roughly the size of a football field.

We had a driver who would taxi us between our houses and the five or six local schools where we were divvied up. As I climbed into the minibus on the morning of my first day, I told him the school I'd been assigned.

'No, no,' he said, 'that's impossible. We pulled all the volunteers out of that one months ago.'

The way he delivered this line, it very much sounded like the following one would be: 'After what happened to the last kid we sent in there … God rest his soul.' But he didn't elaborate, and I was too desperate to play it cool to pester him further. I double-checked my printed schedule and, sure enough, that *was* where I was meant to teach. When I showed it to him, he shrugged, as if to say 'Whatever. It's your life.' As I watched groups of two, three, and four volunteers get dropped off in front of their schools, my apprehension grew. It didn't help that everyone else was in a cheery mood, singing along to a VHS tape of American R & B music videos from the 2000s. I recognized one of the songs – Usher, I think, or Mario – and tried to join in (fake it 'til you make it), but it wasn't long before I had to switch to the women's parts, as all I could muster was a tremulous falsetto. Between songs, I interrogated those who remained in search of a potential co-worker, but this was fruitless. Sure enough, I found myself dumped from the taxi van at its last stop, lone sacrifice to two hundred waiting juveniles.

I was greeted by the head of the school's sports department, Pastor Samuels (but I could just call him 'Pastor') a rotund, goateed, coloured fellow who, I would come to learn, wore a different parachute-nylon tracksuit each day of the week, with matching baseball cap. (That day's, if you're wondering, was dark purple.) He seemed harmless enough, and my anxiety gradually subsided as he gave me a tour of the school facilities and a briefing on what his usual lesson plans were for each group. These invariably consisted of making the kids run laps around the field to warm up before separating them into groups and passing out jump ropes, soccer balls, or pieces of chalk for hopscotch on the concrete. We ended the tour outside his office.

'So,' I asked, 'when's our first class?'

He skimmed over the laminated copy of the timetable I'd been given. 'Let's see. Ah. Here it is. You'll be taking the year eights after lunch.'

'*I* will?'

'Oh! I should have mentioned – I won't be around the rest of the week. I have a few appointments. Here are the keys to the equipment shed. And remember, don't be afraid to tell their teachers if any of the students misbehave.'

'Sure thing, Pastor.'

With this, my stint as adjunct gym teacher began. I would take three or four classes per day, from grades one through eight. Admittedly, it seemed he'd been running quite a regimented operation. To quiet any chattering before the session would start, he had trained the students to respond to his 'Good morning, class!' with a Marine-Corps style, 'Good morning, Pastor Samuels, God-bless-you-to-day!' (He was, after all, also the self-anointed head of the school's religion department.) This was followed by total silence while they stood at attention. I quickly got rid of the second half of the salutation, but I liked eliciting a 'Good morning, Jack!' every once in a while.

Unfortunately, this opening exchange only bought me a few minutes. Almost immediately, I realized that, without a whistle (or, ideally, a *foghorn*) to aid in the wrangling process, I'd be rendered pretty much impotent. This was payback for all the trouble I'd given substitute teachers over the years. As if trying to herd several dozen excitable nine-year-olds from the townships of Cape Town wasn't challenging enough in its own right, my being a blond American made commanding authority nearly impossible: the children were far too preoccupied with trying to touch my hair, which fascinated them.

Apparently, whistles were in short supply in the greater Cape Town area. By the time I'd secured one after two weeks or so, the damage was already done. But the children weren't my only adversary, nor, for that matter, even *close* to the most formidable. That honour went to the elements.

INTO THE WILD

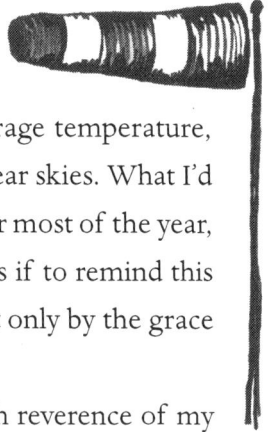

Foolishly, I'd only checked the region's average temperature, which had looked agreeable – warm sun, clear skies. What I'd really needed to consult was a wind sock. For most of the year, gale-force, biblical winds batter the Cape, as if to remind this little spindle of land that it's allowed to exist only by the grace of the gods.

Even so, I struggled to muster up much reverence of my own. Awestruck as I'd like to have been, it was impossible to focus on much beyond my own survival, which felt seriously threatened. Indeed, it was within this dirt quadrangle that the true extent of my frailty would be exposed.

The trials were both physical and psychological. Being left hoarse for the first few weeks from bellowing into the rolling sandstorms was the least of it: the twisters chased and tormented me, their tendrils whipping and scouring; their granules burrowing, weeviling, *probing* into my every orifice, my every nook and cranny. I don't usually do product reviews, but I've got to say my contact lenses, for their part, put up pathetically little resistance. Within minutes, each morning's new pair was irreparably desiccated, surrendering all flexibility and transparency, making it appear to onlookers as though I'd been stricken by matte-grey cataracts. Of course this situation also blurred my vision and left my eyes weeping and blood-shot. After a few days, I gave up and switched to glasses.

But the humiliation didn't end there. My poor old nasal filtration system was so overwhelmed that a sort of respiratory menstruation was triggered: grisly nosebleeds plagued me several times a day as my sinus cavity shed its inner lining. Regrettably, there were no tampons sized to fit my nostrils, but even if there had been, I couldn't risk getting caught raiding the dispenser in the girls' bathroom. (*Can you imagine? 'Please! It's not what it looks like!' 'Do enlighten us. What the hell does it look like?'*) I therefore had to make do by stuffing my nostrils with wads of toilet paper (from the men's), which dangled down, tickling my upper lip as I stomped around barking orders. The single-ply tissue, so cheap it was almost transparent, didn't absorb a drop of my blood, which ran down behind the little paper stalactites before drying in the wind and scabbing over to form a crusty little toothbrush moustache.

I must have made for a pathetic sight by the end of the day: blind, feeble, wheezing, any and all distinguishing features having been sandblasted from my face. I felt like the English Patient, or one of those Peruvian mummies they find up in the Andes, well-preserved but stiff and leathery. I'm sure my students would have been sympathetic, but I resolved to suffer in silence. Putting on a brave face, however, was a bridge too far; the best I could manage was a sort of pained grimace, which left me looking more constipated than stoic.

ટ▲

However fraught my relationship with nature was during work hours, I was happy to shelve our differences in my free time. In pairs or small groups, the other volunteers and I would climb the neighbouring Muizenberg Mountain, or call a taxi to take us further afield to any of the dozens of trail-heads around the peninsula.

The more I explored, the more I began to see the city as an expendable accessory to the peninsula, a landscape so impossibly beautiful that any sign of human civilization – our trappings of steel, brick, and cement – no matter how well designed or aesthetically pleasing, would have been an eyesore. That the city was fated to be overshadowed was true in a literal sense, too: just as water settles into the lowest possible point, assuming whatever form it's allowed by the solid structures that contain it, so behaved the human sprawl, filling what space the topography allowed. The city proper is contained in a carved-out bowl on the peninsula's north side, and loomed over by Table Mountain, whose sweeping plateau looks impossibly level from the streets a thousand metres below.

Almost year-round, even on otherwise cloudless days, nature employs its party trick, creating a thick layer of white fog – a 'tablecloth' – that sits firmly atop the plateau, refusing to venture more than a few metres beyond its comfort zone. Occasionally, a particularly strong wind front yanks a few wispy threads down the slopes and toward the city, conjuring the powdery head of an avalanche that perpetually spills part-way down the slope but is never able to gain ground.

Devil's Peak and Lion's Head, Table Mountain's two semi-attached mountains, bookend the larger one with their own distinct, if slightly lower, peaks, forming the geologic amphitheatre of the City Bowl. Viewed from the north, this trio gives the city an iconic skyline, and its official logo: two deputies standing guard, one, Lion's Head, facing west out to sea and the other overlooking the mainland to the northeast.

The edge of Table Mountain presiding over the western side of the peninsula is much less uniform, though no less imperious, than the flawless crescent to the north. Here, massive ravines cut down through the rock face, dividing the edge of the mountain into jagged, dentate peaks. This magnificent silhouette, confusingly known as the Twelve Apostles, overlooks immaculate beaches as well as the wealthy neighbourhoods that have colonized the narrow tract of land between the mountains and sea. With workable land so scarce, architects often get creative, building around or over boulders and gullies.

Getting here from the city, however, is easier said than done. The long way, around Lion's Head, sends you past the Victoria & Alfred Waterfront, a shopping centre with galleries and restaurants surrounded by heavy traffic. Your alternative is to try Kloof Nek, a narrow pass over the thin, raised bridge of land that keeps the Lion leashed to the Table behind it. This route is often closed due to forest fires on either of its neighbouring slopes or bottlenecked by car accidents. The longest – but certainly the most beautiful – route is the Chapman's

Peak Drive, which looks out over the Atlantic and Hout Bay after cutting across the peninsula just south of Muizenberg. The road is cut into the mountainside, with sheer slopes above and below. An ideal setting to film a thrilling car chase or a commercial, a not-so-ideal one to take your eyes off the road.

However you manage to get here, it quickly becomes clear how Lion's Head got its name. The mountain's base, a wide, grassy hill, slopes gradually upward before producing an incongruous peak of exposed rock that, in profile, resembles a lion in sphinx-like repose gazing out over the water, monitoring the comings and goings. The view from atop any of these mountains confirms that this is a full-time job: a never-ending train of cargo ships chugs along the shipping routes that skirt the continent. Their wide, languid wakes remain long after they've passed, making it impossible to tell whether these are merely aquatic contrails or the very currents they overlap, offering hikers an enticing illusion: to perceive the imperceptible.

The current in question is the Benguela. Waters from the frigid South Atlantic travel up from Antarctica to combine with cold swells that rise from the depths nearer to Africa, hugging its western coast and reaching as far as the equator. Depending on the season, whales drift along this pipeline and, closer to shore, pods of dolphins churn the surrounding waters as they feed on the schools of fish drawn to the nutrient-rich channel. Sharks are also visible on occasion, especially to the east, across False Bay. Rarely, you'll catch a terrifying glimpse of a great white's dorsal fin, although the owners of these are less inclined to approach the shore than their favoured prey, the Cape seals, who are regularly seen fishing and surfing the waves just offshore. Humans, too, are drawn to the water, since the wind accompanying the Benguela happens to create perfect conditions, for not only nosebleeds, but water sports (including, as we'll soon learn, kite surfing).

BONE APPÉTIT

Despite the impossibly majestic natural surroundings, the fact remained that my school still found itself in a rough, underfunded catchment area; a predominantly coloured neighbourhood just a mile or so from a township. For many students, their school-issued uniforms were the nicest clothes they owned, and they'd wear them for weeks on end: a starched white button-down shirt, charcoal-grey wool shorts or long trousers for the boys and pleated skirts for girls. Their shoes were simple, black, Velcro-strapped trainers and their gym clothes were white cotton shirts and shorts. In many cases, school was the safest place these kids could be, especially for those coming in from the townships, which were often violently controlled by gangs, with almost no police presence, and where the HIV/AIDS rates were often over 35 percent. That said, safety remained a concern even here at school, where tall, barbed-wire topped fences lined the property, and the teachers instructed me not to leave the premises at any point in the day; even collecting a ball that had been accidentally kicked out onto the street was forbidden. I took these instructions as suggestions rather than orders, reserving the right to quickly fetch a football if needed. Any rebellious urges I had, however, were soon quashed.

A couple of weeks after I'd arrived, my school closed for its midterm holiday, during which our organization offered a day camp for the students, many of whom had nowhere else to be. It would be run by me and a couple of gym teachers from the bigger school nearby. I'd planned a light itinerary, hoping to pretty much leave the kids to it so I could do some relaxing of my own. With the school off-limits, it turned out that the only place we could entertain several dozen children was an empty lot inside the nearby township. This was quite controversial. We, the volunteers, would be allowed access, but had to be accompanied by a security detail, as we would technically be on someone's 'turf,' and would therefore be at serious risk if we didn't have a respected local there to vouch for us. (We were assured that our students would be fine.)

Not to worry, though; the administrators kept a local man on retainer for just such an occasion: Charles, a forty-something coloured man who had rhinestones inlaid in what few teeth he had left. He was an allegedly reformed gang member-turned-community servant. I'd never met him, but I knew he helped out at a couple of the other schools; the other volunteers had all said he was great to have around. This proved true enough. He loved helping with the kids and also keeping the volunteers abreast of the township's goings-on. On the first day of camp, he informed us that, just last week, men from his neighbourhood had 'dealt with' two young men who'd tried to burgle a family in the township – this was commonly a part of a gang initiation. He implied that the robbers had been beaten either to death or brought very close to it, which was apparently the standard penalty for those determined (by community elders) to have committed any serious crime. In the townships, Charles explained, these kangaroo courts are the only reliable mechanism for any form of swift justice; they provide structure and security in otherwise lawless territory. For what it's worth, he deftly avoided our lines of inquiry regarding the extent of his involvement in the burglars' punishment. It seemed like plausible deniability / omertà were concepts taught comprehensively in Street Justice 101.

Trusting the moral compass of the same system that encouraged 'corrective rape' to cure women of lesbianism felt dubious, but I suspected that we, the nineteen-year-old white kids from Middle America, weren't ideal crusaders for the front line of this particular dispute. In the end, all we did was tell Charles we were glad those robbers were off the street. Moral outrage, it turns out, is much easier behind a keyboard.

While processing Charles's story, I wandered around the edge of the lot, taking care not to wander too close to the border and risk copping a trespassing charge. One of the boys, a first or second grader named Josiah, tagged along. Though we were only a mile or so away from school, I pointed out how different the mountains looked from this new vantage point. He then showed me our school's mascot, the guardian 'elephant'. From where we stood, the nearest portion of the grey ridgeline curved gently before slanting downward, carving out the animal's recognizable sloped forehead against the western sky. Where the forehead overlapped the next slope to the south, the resulting shadow carried on downward to ground level, tracing a low-hanging trunk; a large cave set into the mountain gave the profile a perfectly proportioned left eye.

Like the field at my school, most of the lot's top layer had been stripped back by the wind. In a far corner, I saw a small cluster of third graders on all fours digging at something. Among them was Kiana, a sweet, pigtailed girl who was missing most of her front baby teeth. When she saw me approaching, she ran in my direction carrying what looked like a stick. As she got closer, I realized she was holding what looked, to my untrained eye, to be a human femur. Brandishing the artefact gleefully, she called for my attention.

'Look what we found! We think there's more! Come help us!'

There may well have been more, but we weren't going to find out. I ushered the kids away to an area of the field that I hoped was less likely to be a burial site.

What had started as a wholesome, innocent outing had turned into a crisis of perspective. I winced, remembering the immense distress my dried-out contact lenses had caused me. Maybe the cranky old folks back home were right, my generation *could* use a bit of toughening up. These children, in comparison, compartmentalized the real-world hardships around them with ease, deriving genuine enjoyment from even the simplest pursuits. This isn't to romanticize their situation, by any stretch, but merely an observation that the existence of poverty and disease had not destroyed their capacity for joy or optimism. They were proof, in some small way, that our outlook and frame of reference play a bigger role in our day-to-day mental state than we'd like to admit. This sounds clichéd, I know, and obviously I only saw these kids at their most excited, but I never heard or saw any of them express angst, or cynicism, or displeasure about the state of the world. If they were upset, it was because another kid had taken their basketball or jump rope, or they'd fallen over and cut their leg. That said, they were tough as nails: endearingly – and tell-ingly – every single bodily ailment, from hangnail to shark bite, was described by South Africans simply as 'sore'. They didn't resent me for anything that either of our governments had done before we were born. They would ask me about life back home, peppering me with questions like: 'So, have you ever played a PlayStation?' When I'd say yes, they'd get excited, desperate to hear more. Learning about a cool gadget didn't make them jealous or disappointed like it would have done to me at that age, it just made them more curious, as if I were describing a spaceship or my top-secret superpower. In return, I'd ask them what they did when they went home and heard all about their favourite games and TV shows and sports teams. We'd then go back to playing games or running races.

I'm not sure if Pastor Samuels intentionally leveraged their naïveté and exuberance, but he certainly benefited from it, never more so than when class resumed the week after the holiday. The time off had done the students no good, he declared; they'd already been lapsing toward sedentary life-styles and, with the local youth athletics competition coming up, a rude awakening was in order if the kids hoped to podium in any of the events. I was no expert, but Sloth and Gluttony seemed to be the least of our worries. The kids looked closer to malnourishment than anything else – in most cases, their knobbly knees were the widest point of their legs. Nevertheless, he insisted that the children needed to be 'whipped into shape'. (I took his directive as a figure of speech, though I'm sure I could have enforced it literally as long as his daily siestas had remained uninterrupted.) Football and jump roping – crowd favourites – were swiftly banned; in Samuels's view, they let too many participants get away with standing around doing nothing. Gym classes would now feature only rudimentary (and compulsory) cardio training. With plastic cones the only equipment at my disposal, my only recourse was to make the students do laps of the field behind the school until they were exhausted. Miraculously, the kids didn't harbour any of the bitterness and resentment I would have felt if forced to run relay races around the same dirt patch for two months straight, especially knowing that the baton we were passing could be my late cousin's fibula.

BAFANA BAFANA

I'd soon get the chance to witness the country's enthusiasm for sport on a much larger scale: the national soccer (as association football is called in South Africa) team had a match the next weekend, and my housemate and I decided to go.

We'd bought our tickets from a grocery store, of all places. Given the widespread lack of internet access, these shops were the main ticket outlet for sporting and other public events. As your cashier scans your loaf of bread or vegetables through the conveyor belt, you can ask if there are any available seats for the city's upcoming rugby, soccer, or cricket matches, or even concerts (Rhianna, Eminem, and Justin Bieber had performed in Cape Town within the last year or so) But the offerings don't stop there. You can also ask for phone credit, whereupon the cashier prints out a code you can send to your phone company for, say, thirty minutes of call time, or a few megabytes of data to browse the web. And you can grab a few days' worth of electricity: just give the cashier your address and a few rand, and they'll print off a string of numbers for you to enter into your meter back at home.

My wing-man for the match was a twenty-something Ghanaian from Canada called Aaron; we'd struck up a friendship given our shared interest in sport. Our first interaction had taken place on my first morning after arrival, when I'd awoken to hear laboured breathing coming from the small patio behind the house. I panicked, fearing the knife-wielding robbers were back to collect another haul while we were off at school. But we were safe, for the time being – it turned out to be Aaron, who was doing push-ups on the cement. I went outside to introduce myself and noticed a deck of cards beside him. He did a few more reps, then flipped over another card. A joker. He sat up for a few seconds and, seeing my confusion, explained the rules.

'Whatever number you get, you do that many push-ups. Aces and faces are supposed to be five each, but I usually do ten. Jokers are a fifteen-second rest.'

'Got it. So you just do one or two of the cards per day?'

Not letting on to the stupidity of my question, he graciously explained that, no, he did the full deck every morning. It turned out that he was actually a semi-professional American football player who was just travelling for fun between seasons. This workout, he told me, was the invention of Ray Lewis, a superstar linebacker in the National Football League whose name was synonymous with intimidation and delivering punishment. (Unfortunately, Mr Lewis's predilection for violent scrums was a tough habit to kick, even as a civilian. In 2000, only a few years after his first Super Bowl win, in which, to his credit, he was the Most Valuable Player, he was charged with murder – I hoped Aaron didn't idolize him *too* much.)

He invited me to join him for the rest of the deck but, seeing he was only about a third of the way through, I politely declined. *Ten* push-ups would have been a bit much for me, let alone what looked closer to two hundred. I'd never done much muscle-building; most of my exercise came from playing tennis, which was cardio-intensive and required more lower-body strength. I wasn't scrawny by any means, but felt particularly so next to him. By comparison, my chest looked concave enough to be used as a cereal bowl. At any rate, we decided to take the train into the city on the day of the match. Taking public transport was usually strongly discouraged, but we reckoned we could hold our own, or at least give off that impression: we were both six feet (1.8m) tall and athletic-looking enough. (For good measure, I wore a baggy hoodie to enlarge my silhouette.) The trip went pretty smoothly; the trains, to their credit, ran pretty much on time and were in decent condition, other than the graffiti. I noticed a couple of weird stickers advertising something to do with abortions and psychic readings, but I kept my gaze locked to the floor to avoid any unwanted eye contact.

ॐ

The stadium was situated on the northwest corner of the peninsula, no more than a hundred or so metres from the ocean. The stunning location made access inconvenient, however – from the train station, we still had to trek a few miles to the far side of the city. In the end, we were grateful for this; the pilgrimage felt like an integral part of the experience. The main road through the city had been blocked off to vehicles and was packed with throngs of people who converged on the white, ethereal glow of the stadium in the distance. As we got closer, the crowd swelled, and the klaxon of the vuvuzelas became deafening. Following the World Cup, these controversial plastic trumpets, which had given the tournament its memorable, and distracting, locust drone backing track, had been swiftly banned in stadiums worldwide. Here, though, roadside merchandise stalls still sold them in every size and colour.

Before kick-off, the man next to me told me that he and his friend had saved their money for a couple of weeks to buy tickets (which had cost the equivalent of about eight dollars). But he was happy to treat himself, he said, because of how long he had waited for this opportunity. The stadium, he explained, had been purpose-built for the 2010 World Cup, and had hosted several games, but locals like him had been priced out. In the years since, there still hadn't been a chance to see their beloved national team (which they called Bafana Bafana, meaning The Boys). Because of the national politics behind the sport, they played most of their matches in the heavily populated hubs of Johannesburg and Pretoria. The Western Cape, being affluent and proportionally whiter, preferred rugby (the favoured sport of the apartheid regime), which had its own stadium nearby. This explained a lot; it seemed like everyone here had several years of pent-up energy to burn.

Neither team had much to play for, so it was essentially an exhibition match, finishing as a nil-nil draw with only a few shots on target – uneventful by any standard. Not that you'd have known: entire sections of the stadium seemed to have been reserved for coordinated fan groups with matching face paint and elaborate costumes, who sang and danced joyously in choreographed, rhythmic unison for the entirety of the match. For their part, the fans around us were riveted. There was a hum of anticipation whenever Bafana had the ball, even in unthreatening positions. If a South African player drove the ball forward into the opposition half, the crowd rose in keen anticipation, bouncing on restless feet. Each shot or cross drew shrieks of excitement. Early in the game, one of the players overhit a pass to a winger, putting the ball out for a throw-in deep in Nigeria's half. The man next to me squeezed his eyes shut and threw his head back to release a scream of pained frustration into the sky, reaching for his hair as if to pull it out. Many others did the same. Thankfully, the Nigerians were by no means a ruthless attacking outfit; there were only a few occasions where they threatened the Bafana goal. Had the opposition scored, I estimated there would have been at least a few hundred conniptions suffered amongst the home support.

I found it impossible not to get caught up in the histrionics; not long into the match, every misplaced pass or shot elicited from me a tortured groan just like the crowd's. If a player slowed down an attack by lingering on the ball instead of releasing it to a teammate who'd made a clever forward run, I also threw my hands up in theatrical exasperation. If he at least attempted to play the ball, we politely clapped in encouragement. But heaven forbid he turn and pass backward to the defenders or goalkeeper or, worse still, decided not to slide in and tackle a Nigerian for a loose ball. Upon such displays of cowardice, vitriol cascaded down toward the pitch. The feeling of community and the release of raw emotion en masse was particularly cathartic because of my relationship with tennis, a lonely sport by definition, made more so by its obsession with stiff propriety. Usually, it was only in private that I could bemoan a player's lack of testicular fortitude and/or question his upbringing. Here, though, I had thousands of friends with whom I could vent my frustrations. (Our hatred was, of course, forgotten seconds later as we praised the very same player for intercepting a pass or winning a throw-in.)

By half-time, the two beers I'd had before kick-off had caught up with me, so I ran to the bathroom, along with what looked like half the stadium. The urinals, to my dismay, were of the minimalist variety – just a sheet of aluminium covering the bottom half of the wall, with a trough running along the base toward a drain in the corner. Point being: worst-case scenario, privacy wise. You don't know performance anxiety until you're trying to gee yourself up to use an African urinal under the gaze of hundreds of impatient locals. Things can only get worse if you think too much. Still five or six rows back from your turn at the wall, however, thinking was all there was to do.

As we shuffled shoulder to shoulder toward imminent disaster, I spotted something on the Wailing/Weeing Wall – some *things*, rather: the same ads I'd pretended not to see on the train. There were *hundreds* of them plastered down the length of the room, so densely that they overlapped and abridged each other. Not that the gist wasn't still abundantly clear: PENIS ENLARG PENI LARGE HELP NOW PENIS BIG STRONG ERECTION PEN PE PENI P STRONG EJACULATE DR HELP GIRTHY PROPHET POWERFUL EJACULATE HELP.

No, no, no. Please, God. What had I done to deserve this? It was like some sort of fever dream where my computer had been hacked and my search history was being broadcast for the world to see.

I was up next. I assumed the ready position, feet shoulder-width apart, and stared at my reflection in the wet metal. This was it. The moment of truth. My grandfather (not the affirmative action-dissenter, but the other one) hadn't all but frozen to death in the Battle of the Bulge for *this*. Whip it out and drain the main vein like a man, for Christ's sake. *OK, then. The moment of truth ...*

I unzipped my trousers to retrieve the aforementioned 'vein', but he refused to avail himself, having shrunken in fear to more of a capillary. Just then, there came a roar from the crowd outside; the second half must have been about to start. *Y'know what? Grandpa was a big sports fan – he wouldn't have wanted me to miss a moment of the action. I didn't need to go that badly, anyway.* As I got back to my seat, I saw Aaron, who mentioned he'd gone to the bathroom in the other direction.

'All good?' he asked. 'You took a while.'

'Me? *Oh*, yeah.' I puffed out my cheeks. 'Like a racehorse. Line was kinda long, that's all.'

'Well, you should have gone with me. It was pretty much empty.'

I waved a hand carelessly. I was so relieved to have avoided public humiliation that I was able to forget about my swollen, aching bladder until full time. But there was to be another nightmarish situation before the night was through.

It was almost midnight by the time we got out of the stadium, so the trains were no longer running. We walked back toward the city with the other fans for a mile or so. Our rideshare apps said there were no drivers who could get to us in less than half an hour. But we knew we couldn't afford to stop walking or look even slightly confused or lost, so we kept walking. Alas, the crowd, serviced by ranks of minibuses advertising rides back to the townships, began to thin out, leaving us somewhere on the poorly lit outskirts of the city. Which leads us to our other problem: we had no cash. We'd only brought enough for a couple of drinks each, and only had about five dollars left, which wouldn't have been enough for a normal cab even if there were any around. Withdrawing more was apparently out of the question: I'd brought my debit card, but as we'd walked by a gas station with an ATM we'd overheard a man walking ahead of us warn someone else not to go near it, as a local gang had been using it recently to mug unsuspecting customers.

This meant our only option as far as transport that we could afford was to jump in one of those minibuses (which locals just called taxis). These were the city's equivalent to a subway or metro system. They screamed down the highways and main roads on set routes – mostly from the townships to the city and back – as vital, albeit unsanctioned, public transport. In most cases, there would be a driver, and another guy, called the *gaatjie* (pronounced *haa-chee*, the *h* being guttural), usually in flip-flops or barefoot, who would stand hanging out of the big sliding door on the side, screaming out the taxi's destination, often holding a cardboard sign (for any hard-of-hearing commuters, presumably – although they'd have to be pretty much stone deaf, as you could hear these fellows from halfway up Table Mountain). The gaatjie would alert the driver if anyone flagged them down, which was usually done by waving a green ten-rand note. After collecting the money and making sure everyone was on board, he'd whip the door closed and sit on his upturned milk crate, to emerge in another fifty metres or so. Compared to the train, these taxis ranked only slightly lower on the 'Things You Must Never Do' list as prescribed by our programme coordinators. Even the fellow who dropped us off at school each day and drove one of these taxis for a living had only words of admonition. For what it's worth, the safety concerns relayed to us were less about the road-worthiness of the vehicles, and more due to who was in charge of them. The routes were often managed by gangs from the townships or the Cape Flats who were constantly at war with each other for turf, and with the police over licensing and permissions. Drivers were always being pulled over for transporting far more passengers than the vehicles could safely transport; in one routine traffic stop during my time there, the police found forty-six children packed inside a van designed to seat nine. In fairness, though, in all my time there I heard more stories of crime on trains than on these taxis.

We walked for half an hour through the city, streetlights flickering or shorted out altogether, traffic thinning, beggars and vagrants loitering. We quickened our strides, doing our best to look busy and irritable, a skill I would, in time, come to perfect, but on this night I had a feeling I wasn't selling it. As each minibus passed us by, we listened out for a cry of 'Muizenberg!', but no joy. Nobody was heading out to the almost all-white beach suburb at this hour; all the (black) service staff had been extracted from there hours ago. We approached a gas station, mercifully well lit, where we saw a minibus parked, with a man we assumed to be its driver leaning against the door. We greeted him and asked where he was headed. He wasn't working tonight, he said. But when we offered him everything we had, he looked up from his phone and, after looking us up and down, said he'd take us, and nodded at the sliding door for us to get in.

Before we got moving, he dialled a number excitedly, but after a few rings he seemed to think twice, hanging up and typing out a message to the person instead. He then started the vehicle and we set off. We tried to make conversation, but he seemed distracted and fidgety. When we told him our names and asked for his, he didn't respond. I'd never seen any of these taxis not blaring music until now. He made several phone calls, but spoke only in what sounded like Xhosa. The roads he took weren't any I recognized, they were unlit and, by the feel of things, pretty much unpaved. Remembering the guardian elephant Josiah had shown me, I looked out of the window to try to get my bearings – not that they would have made much difference – but it was too dark, regardless; the comforting glow of the stadium was no longer visible in the rear-view mirror. I looked over at Aaron, sitting next to me in the middle row, who I could tell had the same concern I did: that this ride was going to cost us a lot more than our handful of crumpled dollars. We needed a plan.

At that moment, the driver yanked the steering wheel, skidding us into a parking lot, bringing the taxi to a screeching halt. Taking the keys out of the ignition, he remarked that the engine was broken; we were going to wait for his friend to arrive and take us the rest of the way. Aaron and I looked at each other in silent agreement. I opened the door, explaining that I was just going for a pee behind a nearby bush. Aaron did the same, finding his own spot nearby. The driver agreed, perhaps figuring that we hadn't yet voiced any concern and therefore weren't flight risks; denying us exit would have risked spooking us and giving the game away. As soon as we were a safe distance we started sprinting, safely reaching a gas station we'd passed a few hundred metres back. Our brush with abduction notwithstanding, I considered the evening a triumph.

T.I.A.

Broader societal issues aside, I knew that the thrill of playing outside was simply a law of nature for young kids, one that had nothing to do with me. Not that I was going to let this faze me: being the students' favourite teacher was perfect fodder for my burgeoning martyr complex. My lack of co-workers made my experience different to that of most other volunteers, who could share the load for child-wrangling and lesson-planning, and whose matching schedules meant they often stuck together round the clock.

This was good news for the coffers of the local pubs, as volunteers swarmed to their karaoke nights and happy hours throughout the week. Harmless fun, one might think, but unfortunately a pattern developed for more than a few volunteers, as staying out all night for Karaoke Wednesdays began to take precedence over showing up to school on Thursdays. When the programme coordinators got wind of this, they sent out a memo reminding us that many of the schools and orphanages depended heavily on our help, so taking a day off to nurse a hangover could mean there would be nobody to prepare breakfast and lunch for a few dozen hungry toddlers. When further complaints began to trickle in from the *actual* teachers at these schools and orphanages, the administrators realized they had been too lenient with their wording: the volunteers had stopped skipping work but were now turning up to school smelling so strongly of alcohol that even the children had noticed. A second, sterner memo was circulated immediately. The volunteers complied, but more for the kids' sake than out of respect for the admin team. Unfortunately, growing concerns around the administrators' trustworthiness had fostered a sense of rebellious self-government among the volunteers. (There were whispers of higher-ups pocketing chunks from our fees, which were supposed to go solely toward our room and board.) This meant the group became a sort of sovereign microstate, replete with its own social hierarchy, norms, and even its own lingo.

It wasn't long before I noticed that a curious phenomenon had developed, whereby volunteers justified any hedonistic excess by explaining that this was a 'once-in-a-lifetime' trip. In the same vein, the aphorism 'T-I-A!' (This Is Africa!) became popular. Obviously, the factors at play here – cheap booze, nice weather, free time – were by no means endemic to the region, but that didn't seem to matter. It was used as a response to any hardships or inconveniences faced, as well as a catch-all rationalization for anything else. I'd heard (and used) many excuses in my time, but this was the first I'd heard anyone blame an entire continent.

At any rate, the three remaining members of the Soul Sisterhood decided the acronym so perfectly summarized their trip that the only appropriate course of action was to get matching *TIA* tattoos. I tried to point out that the phrase didn't necessarily carry a positive connotation, but my concerns went unheeded; the appointment was made official. After seeing the girls off, the rest of us waited anxiously, killing time by placing bets on where their tattoos would be. (Given the demographic, the smart money was between ankle, wrist, or side-boob.) They were back in under an hour – they only needed nine letters, after all. But when they revealed their new tattoos, it became immediately clear that something had been lost in translation (looking back, this should have come as no surprise, as South Africa has almost a dozen official languages). Instead of 'TIA', there was 'Tia'. *Tia* translates to 'aunt' in Spanish, meaning the girls now appeared to be a trio of bereaved nieces with tributes to a beloved – yet mysteriously *nameless* – woman etched indelibly into their skin. With their flights booked for the next day, however, it was too late to make any alterations.

Though the girls' wounds were still too raw for the irony to be appreciated, I would argue they had managed, unintentionally, to perfectly capture the spirit of the phrase they so badly wanted on their bodies: T really *was* A.

᠕

I began to observe a niche forming adjacent to our little ecosphere: in the same way that the local criminals had begun to take advantage of our predictability, so too had the local lotharios. It seemed that every bar and restaurant nearby was frequented by several men intent on seducing the young female volunteers, ostensibly offering some combination of excitement, danger, and authentic African experience.

One of these was lanky, coloured, and known for his recognizable afro, but even more so for his rendition of Shaggy's It Wasn't Me, which was apparently so impressive it was the only song he needed to have in his locker. Like clockwork, our afroed one-trick pony would dutifully attend every Karaoke Wednesday. He'd arrive as soon as the stage opened for singers (or rappers), do his three-minute act for the volunteers (who made up 90 percent of the audience), then descend into the now-raucous crowd to schmooze and enjoy free drinks from his starstruck groupies for the remainder of the night. One girl from New Zealand succumbed to his siren song (or, technically, his siren *rap*) and actually cancelled her flight back home to pursue her infatuation. Sadly, their fling only lasted a few weeks before his song's lyrics proved prescient: his honey, if you can believe it, caught him 'red-handed, creepin' with the [volunteer] next door'.

There was another guy with an afro, although this one (the guy, that is) was white. A mellow, slightly scruffy kid, he went by the name of Poseidon. While his modus operandi was arguably less refined than the rapper's, it gave him more regular access to the group. His meal ticket came from bringing over a handful of skunk weed to the main house most evenings to 'hot-box' their living room. From what I could tell, there had been a non-compete agreement, and he was the volunteers' exclusive source of contraband. He must have made a killing, as there were more than a few volunteers whom I'd never seen sober.

I was living in a small bungalow across town along with a few other volunteers, but stopped over at the main house one Friday after school on my way to the town's weekly food market, and by chance I witnessed the unveiling of Poseidon's new shisha pipe. I'd never smoked from one before and made a fool of myself by knocking it over, *twice*, in the span of ninety seconds, spilling hot coals into the laps of those sitting next to me. In retrospect, I'll take some credit for admirable wingman work, as my buffoonery allowed Poseidon to be the hero, reacting quickly to save the girls to his left and right from disaster. He deftly salvaged the glowing embers before showing us his trick to diffuse the heat from his fingertips:

'Bunsen, as in *Robert* Bunsen, the guy who invented the burner, discovered this. Our earlobes have a unique chemical composition – they suck the heat from your skin if you've burned yourself. You just need to pinch them. Watch this.'

The group swallowed this story whole, each of them testing his theory, even if they hadn't picked up any embers themselves: '*Wow, he's right! I can really feel it absorbing the warmth!*'

After an hour or so, the novelty of his new toy wore off, and one of Poseidon's usual clients asked him if he had anything more exciting on offer. He did, he said with a smirk, but if anyone wanted to give it a go, they'd need to come with him back to his headquarters. I didn't want to embarrass myself any further by being the square who was scared to partake, so I volunteered to join the expedition party of about a dozen people. In the wake of Poseidon, we wove through the back alleys of Muizenberg before arriving at his parents' house where, we were assured, nobody would be home. He pushed the front door open and told us to follow him down a narrow hallway. It was poorly lit, with only a handful of small, flickering, gas-lantern sconces along the odd burnt-orange walls. But even more off-putting was the décor – covering every inch of the walls were antique weapons and what looked like medieval torture devices, along with all sorts of skulls, antlers, and taxidermic animals.

'You guys do a lot of hunting?' I asked our host.

'Uh, no. Not really.' He looked bewildered, as if the question were a non sequitur.

We finally made it to his small, low-ceilinged bedroom, which quickly became cramped and claustrophobic as our party filed in. He pulled a duffel bag full of contraband from under his bed, rummaging through it for a few seconds before producing a small eye-drop bottle.

'Mushroom extract. All you need to do is put some under your tongue, and your adventure begins.'

He held it in cupped hands with great reverence, as if inviting us to behold some sort of mythical elixir. A murmur of excitement spread through the group. I, however, was too distracted by what I'd just seen on the ceiling: an antique bear trap waiting open above me. The family had mounted several of these throughout the house not as kitschy chandeliers, suspended and strung with bulbs or candles (which would have been forgivable, albeit only barely). No, these traps, rusted and decaying, had been bolted directly to the plaster, steel jaws agape, primed to spring into action.

I was (and still am) no expert on hallucination, but if there were *any* power of suggestion element to it, whereby one's surroundings influenced the drug's effects, I suspected that I'd be giving my vulnerable mind the worst possible mise en scène from which to start its adventure. But Poseidon insisted, and, in order not to antagonize our host, I accepted – albeit only the smallest possible dose. Once we'd all received our droplets of the bitter liquid and were waiting for them to kick in, I suggested that we head down to the beach and stargaze. With any luck, we'd have one of those spiritual, cosmic awakenings Poseidon had advertised and so often alluded to. My *actual* motive, of course, was that I'd rather have been anywhere but sitting in his bedroom waiting for the tatty, mangled antelope head to gain sentience and start recounting to us its life story, or worse, its *end-of-life* story …

Looking down upon us in our intoxicated paralysis, the animal's glass eyes would light up. The day of retribution had arrived, and he wasn't going to pass it up. He might start by tearfully describing to his utterly captive audience what it felt like to take a bullet through the spinal cord, the impact of which had shattered several vertebrae, instantly rendering

him paraplegic. He would then describe the harrowing experience of dragging his lifeless back legs through the savannah for miles while haemorrhaging a quart of blood a minute, growing increasingly disoriented and exhausted. His children and partner, who had bolted as soon as they'd heard the gunshot, could only watch on helplessly. (He would make sure to mention that his specific branch of the antelope species mates for life, with the bereaved dying of heartbreak soon after.)

After several hours tracking the animal's blood trail through the brush, the hunter, a twice-divorced, retired dentist who'd paid nearly twenty thousand dollars for this trophy hunt (flights not included) would have finally stumbled upon his prey. Standing over the crippled bovid, the man ceremoniously unsheathes the two-foot long tribal machete his buddies had gifted him before the trip, ready to deliver the coup de grâce. Only then does he realize that he'd completely forgotten to sharpen the ornamental blade. *Ah, fuck it,* he thinks. *The thing's almost dead anyway; it won't know the difference. Besides, animals can't even feel pain.* He would then put the poor beast out of its misery – eventually – after twenty minutes of sawing at its neck. This would have butchered the trophy in the process, of course, rendering it worthless and relegating it to being pawned off to a family friend – the father of Poseidon.

Mercifully, things never took this turn, as my beach proposal was well received. In the end, I don't think I'd taken a big enough dose of the mushroom extract to experience its intended effects, but this may have been for the best as the beaches in South Africa can be hotspots for crime, especially at night. Two members of the group stayed there after the rest of us returned and fell asleep in the sand, only to wake up with their belongings stolen. I never saw Poseidon again (I did my best to steer clear), but can only assume he's still in town, working his magic on fresh-faced explorers, recruiting them for his high-risk/low-budget ayahuasca retreats.

$\approx$

Another local opportunist enlisted the help of his own animal. This one was alive, albeit only slightly more mobile. One afternoon, a few of us decided to go to the local shopping mall. As we discussed logistics, two of the newer girls chimed in to say they knew a driver whom they'd already called upon a few times. Their phones were out of minutes, however, so I offered to call him with mine. I dialled his number as they read it out, and asked them his name. One of them piped up.

'Oh, right, sorry. It's, uh, Mr Lollipop.'

I hung up immediately.

'Excuse me?'

'He's Mr Lollipop, and he drives the Lollipop Cab! I know, it's kind of weird. We asked him about it last time. He said that there are so many shady taxi drivers out there who mug or kidnap their customers, and he wanted to be seen as someone trustworthy and welcoming. And what does everyone like? Lollipops! It's even what he named his dog. This way, he starts out on a good note with everybody he picks up. Think about it, it kind of makes sense, right? You like candy, don't you?'

For the record, *yes*, I like candy. But what I like even *more* is not being molested. And, in my estimation, trying to gain trust by alluding to candy is the first port of call for someone with a sinister motive. Even if you *are* innocent, trying too hard to reassure and ingratiate before even attempting to make a normal impression doesn't help your cause. It's like being five minutes into a first date before reaching across the table and squeezing their hand comfortingly, then explaining that they should 'fret not – the air holes drilled in the boot of my car do *not* indicate that I've planned for you to be locked therein'. Point being: sometimes less is more. The only people who could *possibly* think the candy thing was a clever marketing ploy were those who had never heard of the trope. My feeling, however, was that the existence of such a group had to be a statistical impossibility. This suggested the existence of a fascinating, albeit far more unsavoury demographic: van drivers who *were* creeps, but who'd been successful enough that they could completely ignore cultural pressures. They had stared, unblinking, into the face of modern society, firmly refusing the norms it sought to impose (for example, that kidnapping is bad).

No, no, no, *no*! these girls assured me. The Candy Man wasn't a pervert, he was a decent man. I asked if they knew his name.

'Yeah, we already told you, Mr Lollipop!'

'Yeah, no. His *real* name.'

They didn't, it turned out. But 'why did that matter?' when, after all, 'He brings little Lollipop everywhere he goes! Would a criminal really drive around with a lapdog?'

It *did* seem unlikely, I'd grant them that, but having a pet was no guarantee of one's moral scruples. Just think: Mafia dons always had cats around, and that didn't stop those guys from misbehaving.

Alas, my vote was outnumbered, and we summoned the mysterious man who had named himself after his pet. When he arrived, I saw that he was heavyset and middle-aged. Thankfully, his van had windows and was far less rusty than I'd expected. As we piled in, we saw a small, white terrier lying on a small blanket across the man's lap. One of the girls (a Lollipop virgin like me), was enamoured.

'Oh my God, she's so cute! What a little princess. Do you take her everywhere?'

'Yes, she's my baby. She's paralysed from the middle of her back down, you see, so riding in the car gives her a chance to see the world. You can pet her if you want – she loves girls like you.'

I'll bet she takes after her daddy, I thought; his invitation was clearly a ploy to get the girls to reach into the danger zone, and I was about to suggest as much when he began to release a plaintive, high-pitched whimper. *What the hell had he been doing under that blanket?* I feared the worst. I then noticed the poor creature's wet, beady eyes looking up at her eponymous chauffeur, her lifeless back legs dangling off the side of his lap, and realized the noise was coming from her. It was tough not to feel a pang of sympathy; I had half a mind to reach up and pet them both in support.

As we approached the city, one of the girls noticed a road-side billboard for a shampoo brand and mentioned how long her hair had grown since arriving in Africa months before; she'd been too busy to find a salon. Mr Lollipop, whose grey hair was shaved into a military flat-top, piped up.

'Ha! I *wish* I could grow mine out! I'd look silly with anything else. All of us in the military had to get the same "Dutch" haircut for decades – it's how you can recognize us Afrikaners!'

Considering the spotty human rights record of the apartheid government, we didn't ask him to elaborate on his military exploits, and instead steered the conversation back to safer waters, namely, the *crippled* Lollipop's favourite snacks and television programmes.

To his credit, the man spoke glowingly about the country's new chapter, and seemed keen to play his part in building an inclusive, multicultural society. This wasn't a common sentiment within his demographic. It's worth mentioning that by insulating and putting the white minority on a pedestal, apartheid did them no favours either. With no succession plan, the regime's collapse left a cohort of disgraced, abortive heirs. Whatever promises Afrikaners had been given about a secure, prosperous future for themselves and their families were not to be realized. On top of this, they were tasked to integrate with a majority that resented them. This process would have been even tougher for folks like Mr Lollipop, who had been drafted – many of them involuntarily – into the military to help enforce the totalitarian regime, either on the domestic front or in places like Angola. Even the US, a proud military nation to its core, had trouble supporting the millions of young men who'd fought to defend its interests. Veterans had far higher rates of PTSD, suicide, alcoholism, and depression than the national average. 'Cry us a river', some might say. Nevertheless, South Africa's white minority still sits on a significant chunk of the country's capital and is responsible for nearly the entire agricultural industry. So, for the sake of triage (both societal and political), it's probably worth considering the extent to which *all* relevant groups have suffered.

YOU COULD BE
SO MUCH COOLER

While Lollipop, Poseidon, & Co. loitered on the outskirts, the real threats came from within our ranks. After I'd been there for a couple of months, the administrators reshuffled us to accommodate an influx of new volunteers. I was to move just down the road to a small flat that had just become available. It was owned by a local family who'd built a second story above their garage. They had a small pool which we were welcome to use, and the weather was starting to get warmer, so the switch was fine with me.

Aaron had gone back to Canada, so it would also be a fresh start in terms of room-mates. One of them, a man-bunned guy named Chet, had just arrived. When I got back from class on Friday afternoon, I introduced myself and explained what had brought me here. He didn't approve:

'Gap year? You left school? Fuck that – I loved college.'

'Where'd you go?'

'NYU.'

'Wow, nice. Isn't that like seventy grand a year?'

'Yeah, but I didn't have to pay. Dad's on the board of directors. No chance I would've gotten in otherwise, ha. I nearly flunked out like five times. It was amazing. We'd have warehouse raves going for two days straight. Just going fuck-ing *wild*. Tripping *balls*.'

'Sounds pretty cool.'

'No, you don't get it. We were on the cutting edge – what music do you listen to?'

Before I could respond, he ploughed on. 'All the shit that's popular right now, Kanye, Drake, all those guys. We knew about them *years* before anyone else.'

I gave him the awestruck expression he was looking for, then asked what had brought him to South Africa.

'Yeah, so, after college my uncle pulled some strings and got me an in at J.P. Morgan. Wall Street's fun, but I got burned out after a couple years. Decided to take a break and focus on my own shit. Kite surfing, mainly. This is one of the best spots in the world. Signing up for this programme was easier than finding my own apartment down here.' (I was sure the orphans would be happy to hear this.)

'Whoa, kite surfing? Nice.'

'Yeah, I've been doing it for a while.'

'I'd like to try it sometime.'

'Well, it's an expensive hobby, so not many people can pick it up. I've sunk thousands into it over the years. Six figures, probably. My board alone was a couple of grand,' – he gestured to the conspicuously shaped luggage leaning against the wall – 'but, yeah. Once you get good enough, it's a blast. So, anyway, what's it like here? What's the deal with the girls?'

I told him there was a decent crop, although they'd be ten or twelve years his junior. He took the news well: 'Oh, *fuck* yeah. I'm looking for a change of pace. My girl back in New York is sexy, but you know what they say: "For every hot girl, there's a guy out there somewhere who's tired of her bullshit." So that's kind of where I'm at. I'll marry her eventually, but I have some stuff I've gotta get out of my system first. She's chill, though. She gets it.'

I shuddered for the sake of the girls who would most likely be on the receiving end of whatever 'stuff' he needed to evacuate from his system.

As usual though, I spared a thought for myself – if he *was* able to convince these girls to sleep with him who could know the extent of his sexual proclivities once he got them into his bunk? I had a hunch that vanilla stuff wasn't going to suffice. It was a tiny apartment, so I wouldn't have plausible deniability – let alone privacy – regarding what might transpire. With the wall between us so desperately thin, would sound and smell travel with such high fidelity that a stimulation threshold might be crossed, and I'd qualify as a participant? If so, would I be jointly liable to pay any child support?

Whether this was a cocaine-fuelled spree of infidelity or a Genghis Khan-inspired mass insemination scheme, our man's modus operandi soon became clear: you can't win the lotto if you don't buy a ticket, and his plan was to buy as many as necessary. Tickets, in this case, took the form of rounds of shots and cocktails for the girls he was courting. His trust fund and the fact that he'd started working on Wall Street while his competitors (and prospective mates) were still in diapers, resulted in Chet having far more financial clout than the field, an advantage which, regrettably, he knew how to leverage. He also quickly realized that this hunting ground would not punish failure: he could go all out in pursuit of multiple targets at once, and anyone who wasn't interested would leave before getting the chance to sully his good name to their replacements.

Sure enough, within his very first week he persuaded a young woman to go home with him – through the wafer-thin wall I heard several minutes of laboured grunting, which I prayed was his, before she scuttled out of the apartment and back home. The next morning, when I went to the bathroom to brush my teeth, I saw a used condom draped over the toilet seat, with its wrapper floating in the water. Chet was sitting in the kitchen, sipping a mug of coffee.

'Hey, man,' I said, pointing to the toilet. 'That wouldn't be yours, would it?' (I sure hoped it was; there were only two of us living in the apartment.)

'Oh. My bad,' he said, grabbing the soggy latex and fishing the wrapper out of the water. 'I tried to flush it last night but couldn't find the light switch. What a night, though, man. This trip's gonna be fucking awesome. I'm gonna make a killing. What about you, by the way? How many have you hooked up with so far?'

'None yet. I haven't been here very long. There are a couple of girls I'd like to talk to, but it's tricky. There's always a big group around, so you never get to chat to anybody one-on-one.'

'Dude, fuck that. Quit being a pussy! You're a good-looking kid. You could get any of these girls. Look at me, I'm no model. No six-pack. Big nose – oh, and it's not a Jew nose by the way. Confusing, I know, 'cause I *am* Jewish. But it's actually an *English* nose. Common misconception. Look it up.'

Roman nose I'd heard of, but never English. Regardless, it wasn't a minefield I wanted to explore. Thankfully, he circled back to his original point (to wit: getting laid). 'I'm gonna be off the market in a few years. There's no reason you couldn't take my place. You just gotta set your mind to it and put in the time. It doesn't even matter what you're packing. I mean, yeah, I'm fortunate enough to be well-endowed, but I have a bunch of friends who aren't, and they've still gotten with hundreds of chicks. I'm telling you, dude, anyone can do it.'

'Thanks, man. That's really good to hear.'

'Come party with me one of these nights. Watch and learn.'

'Yeah, maybe. We'll see. I like to take my time with this kind of thing. It's draining to go out every night.'

'Come on, man, your attitude sucks. You could be so much cooler.'

'You're probably right,' I said, and started getting ready for school.

FIELD TRIP

After Chet had been there for a few weeks, I was somehow deemed cool enough to make the cut for a week-long road trip he organized along the Garden Route, the country's southern coast known for its stunning landscape and sea views.

Even just accessing the mainland beyond the Cape peninsula is perilous and melodramatic, making clear to prospective adventurers that what lies beyond is a territory of its own. The Cape Town region, with all its cosmopolitan attractions, is a curated, well-worn circuit, you can feel and see that *true* wilderness has largely been beaten back. However, this is not the case in the rest of the country, and it doesn't take long for this to become palpable. You can feel the tension build with every kilometre you put between yourself and the city.

You must first cross through the Cape Flats, a smoggy, featureless lowland, itself a geographic incongruence given the striking topography to its east and west. It's not all township – but it's bleak nonetheless, with a low sprawl of industrial parks and low-income housing, crumbling, graffitied walls, coils of rusted barbed wire, tyre-less cars propped up on cinder blocks. The highway cutting through this purgatory offers an exit to the airport, but otherwise does not seem to encourage deviation. Drifters and hawkers wait alongside the road as lone operators or in small teams, with handfuls of aluminium trinkets or mesh bags of (likely stolen) fruit slung over their shoulders. They race each other to your window at each traffic light – of which there's an excruciating overabundance – all red, of course, and separated by no more than fifty metres. When you finally break free, the road gradually inclines in a sweeping curve, bidding a prolonged farewell to the Cape as it builds momentum, before slingshotting you over Sir Lowry's Pass and into the beyond. Right before you cross over, though, you have a view back across the flats and False Bay, giving you the full scope of the peninsula and its mountains.

From a satellite view, the Cape looks extraneous to the mainland, almost alien. This gnarled, spindly appendix, tacked onto the bottom of such an immense landmass, seems like it could only be the result of a mythical party game on a tectonic scale: Pin the Tail on the Continent, perhaps.

However it got here, this little outcrop, replete with its own biosphere and climate, has well earned its squatter's rights: Table Mountain (as every local tour guide worth their salt will remind you) is older than Mount Everest. Indeed, the Table and the region's other mountains inland are products of Gondwana, the supercontinent that comprised what is now Africa, the Middle East, and most of the other subequatorial landmasses. The mountains we see today are just remnants of the *troughs* of the great ranges that once dominated the landscape. For now, Table Mountain clings on, a barnacle small enough to have been overlooked so far, as one of the official wonders of the natural world. Eventually, of course, it too will be sanded away.

It's really no wonder why we flock to these captivating spectacles – Niagara Falls, the Grand Canyon, the colossal waves off the coast of Portugal – it's a primal pull. Their ineffable power and scale work in conjunction, encouraging, indeed *forcing*, us to confront the vastness of time and space; the immensity of the cosmos. They let us dabble in spirituality without the suspension of rational thought, by reminding us that the sublime and transcendent exist right here on earth. They allow us to us to forget, just for a moment, the triviality; the *absurdity* of the human condition. The million-dollar question, then, remains: how can we hold on to our newfound perspective once we've climbed down the mountain and resumed our normal lives?

Lying in the hostel room's lower bunk on the first night of our road trip, I wrestled with these existential dilemmas. My more pressing concern, however, was just a few feet above me, where Chet was 'balls deep' in an exploration of his own, albeit of a more, well, *tangible* subject: the freshman girl he'd talked into joining us.

&

Chet was a red-wine aficionado, so our first stop had been at a vineyard outside the city where, between us, we'd finished a few bottles by the late afternoon. The bulk of it had been split between him and the poor young woman, who hadn't yet realized that his eagerness to recruit her for the trip had mostly been about procuring a concubine for the road. He'd even begun accidentally referring to her by the wrong name, in spite of our repeated, increasingly less-subtle attempts to correct him. She didn't seem to mind too much, however, and assumed he was joking around.

As the night went on, the revelry continued and we ended up at a wine bar near the hostel. I'd had a couple of glasses with them but had also been talked into trying some of Poseidon's skunk weed that had been brought along by the remaining member of our group: a young, nervous Australian named Teddy.

The main reason Teddy was nervous that night was because he had met someone online and was due to meet them in person at the wine bar – it would be his first date with a member of the same sex. Well, *technically*, it would be his *second* date, if we count my role-playing as the smooth-talking provocateur at the bar to help him prepare. Obviously, there was something of a ceiling on how much sexual tension I could help us muster up, so I wasn't too much help, but fortunately my partner was preoccupied, being far more concerned with preparing himself *physically* – in case he and his date, who was picking him up in half an hour, decided to consummate the occasion. He hadn't wanted to leave things to chance, and confided to me that he was banking heavily on the weed's muscle-relaxing properties to kick in. The strain was called Swazi, as in Swaziland, where it had allegedly been harvested. Despite Poseidon's marketing efforts, the exotic name wasn't an indicator of its quality. The weed, cut with its grimy seeds and stems, combined with the

cheap wine we were drinking, made for a nauseating cocktail – for me at least: I'd ended up needing to throw up fifteen
minutes after my first hit. I retreated to our shared hostel room
and collapsed into my bunk, where I lay for a few hours in the
dark, too nauseated to fall asleep.

Sometime after midnight, I heard the lovers (the straight
ones) stumble in. They clambered into the top bunk and had
either not noticed me, assumed I was sleeping, or (most likely)
didn't care either way. Briefly – all too briefly – it seemed their
momentum had stalled, or even sputtered out entirely as they
settled into a still silence, and for a few fleeting moments I
thought I might be spared. *Perhaps one of them had blacked out;
it sure had been a lot of Merlot ...*

Sadly, this lull was only a period of strategic assessment,
the drawback of water before the tsunami. Presently, the
couple began to engage in what sounded like an attempt at
tender foreplay. *Who said romance was dead?* This only lasted
for a few seconds, however, and once the breathing patterns of
the lovers changed and Chet's signature, warthog-style grunting began, I knew it was too late for action. I'd made my bed,
so to speak, and my only recourse was to bite the pillow and
wait for the storm to pass. (From the sounds filtering down
through the mattress, it sounded like she was having to do the
same thing.) And it *did* pass, ultimately, as evidenced by his
pained, gurgling groan, along with the simultaneous convul-
sion of their mattress. Within seconds, he fell into a comatose
slumber. Upon realizing that her lover had no intention of

sharing the bunk with her for the remainder of the night, his lady friend climbed down and slunk off to her own room. Thankfully, I managed to fall asleep despite the cavernous snores reverberating down through the bed frame.

The next morning, I queried him about his snoring's impressive volume and its unique, guttural bass, which wasn't dissimilar, I'd noticed, to that of his copulation grunts. He chuckled.

'Oh yeah, should've warned you. Whenever I have red wine, especially a good vintage, I snore loud as shit. Even louder if I get laid. But I dunno, I've always felt like it isn't on me to do anything about it. I'm already asleep at that point, so if you wake me up, you'll just piss me off, y'know? So what's the point?'

Perhaps I was still under the influence of last night's Swazi, but he was starting to make sense. Who was *I* to ruin his night's sleep just because I was too much of a quivering pussy? In any case, he assured me it wouldn't be happening again. Not the snoring, that is, but 'relations' with his chosen girl. She'd already begun to get on his nerves, he explained, having not reacted well to the revelation that he was just on sabbatical from his 'smoke-show' girlfriend back home. The scorned lover called a taxi back to Muizenberg, and we never saw her again, as she brought her flight forward a week and left before we got back. (A suitable outcome for us all, I think.) After seeing her off from the hostel, Chet gave me another of his trademark nuggets of wisdom: 'Take my word for it, dude, *never* stick your dick in crazy.'

We drove onward later that morning, stopping first to pick up Teddy, who'd ended up staying over at his date's house. The stranger had been much older and fatter (and shorter, and balder) than his pictures had suggested, but he *had* paid for their drinks, which apparently cancelled everything else out. Plus, given that Teddy had already prepared himself to be taken up the [Khyber P]ass, he'd decided it would have been wasteful not to make the most of the evening in that regard. So, with our full trio aboard (and all varying degrees of chipper), we set off east along the coast toward Mossel Bay.

To our right was the vastness of the Atlantic. This becomes the Indian Ocean beyond the continent's southernmost point, Cape Agulhas, although we discovered that, contrary to some claims we'd seen online, there's no visible demarcation, which was disappointing. On the left, meanwhile, rolling hills carpeted by bright-yellow mustard and soft lavender stretched back to snow-capped mountain ranges on the horizon. Further east, these more cultivated fields give way to lush, untouched temperate forest.

You continue on like this for hours, taking in the great expanses of field, hill, mountain, and sea, all shown in their full, gleaming splendour by the sun, which hung high in the cloudless spring sky. Several towns and hamlets are dotted along the coast, each with their own claims to fame, but even the stretches of uninhabited, seemingly impenetrable wilderness between them refuse to be ignored. Indeed, it's here that the route offers more concentrated doses of exhilaration. One farm, for example, offers the chance to actually *ride* one of the ostriches you see as you drive past the property. Other local companies offer spelunking, river-rafting, and zip-lining. Our first contact with the wild, however, would take place the next day.

THREE MEN IN A CAGE
(TO SAY NOTHING OF THE SHARK!)

After the less-than-welcoming welcome presentation at the dive company's headquarters, we boarded the dinghy and set sail. After a twenty-minute ride, we dropped anchor about fifty metres from 'Shark' Island. Here, the crew tipped the rickety cage overboard and began chumming the water while the captain prepared the tuna head.

As the roof hung open, they asked which group wanted to go first. One of the families on board seemed keen and volunteered. They zipped up their wetsuits, strapped on their masks and queued to clamber down into the cage.

Our group watched on for half an hour or so as the others went. They'd all re-emerged seeming relatively satisfied, but we hadn't seen the sharks too clearly from the deck, where they'd appeared only as flashes of shadow beneath the water. We decided to go last, figuring that it was better to wait for the sharks to get as comfortable with our presence as possible. When the time came, we lowered ourselves into the water. If we had had any doubts about whether the comical thickness of our wetsuits was really necessary, these were dispelled as soon as any exposed skin made contact with the bitingly cold water.

For the most part, the sharks seemed amenable to the tuna-head plan, and swam parallel to the boat, which let us appreciate their scale and their raw power as they muscled their way to – and *through* – their target. The current tuna (the fourth iteration), with his mouth hanging open and lifeless, glazed eyes, served as a gruesome bellwether of life in the open water. As the crew pulled in the rope, his head would frequently get wedged in the cage.

On this occasion, he and I were locked face to face. By the looks of things, the previous foray had gone relatively well, only costing him a small chunk of flesh from his forehead (as well as half an eyeball). Shreds of his severed neck and oesophagus hung loosely, swaying gently in the current. But his reprieve would soon be cut short. The yanking from above had intensified, the crew now working like a tug-of-war team to retrieve his head which, once reeled in, would be hoisted and swung around like a lasso – five, six, seven nauseating rotations – and flung back out to sea. Had his black, glassy eyes been functioning, they might have noticed the tub filled with half a dozen more heads like his own, ready for when the sharks eventually shredded or made off with his entirely.

Once we'd seen several of the sharks' passes, I decided I'd try something new. I'd been filming with my waterproof sports camera until now, but I wanted to capture the moment with some artistic flair. The plan was to time my next video so that, when the captain spotted a shark and told us to duck underwater, I'd turn back toward the camera, capturing my face in the foreground with the shark spanning across the background – a shark selfie, a rare trophy indeed. The animal would need to cooperate by getting nice and close; they were difficult to spot even in ideal conditions, and the underwater visibility here was terrible. The crew's continual dumping of the soupy fish intestines had done little to help matters, turning the water around us a murky, faecal brown. Also, it looked as if the local seals had adapted to have some sort of emergency smoke screen, similar to the escape mechanism of an octopus, but with diarrhoea instead of ink. Nevertheless, one of the crew managed to see a shark approaching.

'Down!'

I forced myself beneath the surface, spun to face the boat, and held out my camera while I grinned and waved at the lens, praying a shark was visible in the background so my video wouldn't leave me looking like a halfwit. This, however, is *exactly* what happened, in a roundabout way. I heard a thunderous, crunching clank, and felt the cage and boat reverberate. The crew must've hooked the cage up to the pulley system without warning. I couldn't believe their impatience. *And before I'd even gotten my selfie!* I looked to my left to see if the rest of the group shared my annoyance. Instead of Chet's face, however, all I saw was a dark grey traffic cone. It took me a split second to realize our fourth wall had been broken – shattered – by a probing great white. I was struck by just how far into the cage it had intruded, well beyond its black, glassy eyes, perhaps half a metre. My indignation toward the crew now became high-pitched shrieking for them to 'Please, please, get us (the fuck) out!', followed by profuse apologizing for ever suspecting they didn't have our best interests at heart.

Only once I was back on land and wrapped in a towel could I appreciate what had happened. We huddled around the communal computer back at the hostel and watched the video dozens of times in succession, pausing it frame by frame. In a way, I'm glad I picked *that* moment to not look, so I didn't have to watch my life flash before my eyes upon seeing a killing machine storm up from the depths with impeccable precision, its sole intent being to mutilate each of us in turn. There was more shark in this selfie than I could – or *would* – have ever asked for. Terrifyingly, thanks to its camouflage and the murky water, it was only visible for a split second before entering the cage, offering a stark reminder that we're never far from being exposed as comically defenceless, no matter how comfortable we may be in our climate-controlled little bubbles.

Somehow, our day was not yet over – far from it, because the others had decided to bungy jump that very afternoon. There had been a scheduling mix-up, and today was their only opportunity. If I hadn't planned on joining them *before* almost losing an arm, I now had the chance to reconsider.

FREE FALLIN'

As we drove east from Mossel Bay toward Knysna and Plettenberg Bay, the landscape grew wilder. It was as if great mountains from the north had arrived at the coast but, in their gargantuan, straining effort, had split into fissures. Forested valleys ran perpendicular to the coastline, while the road, punctuated by bridges, ran parallel. It was difficult to gauge the depth of these valleys, as they were filled by glaciers of fog that reached almost up to the bridges themselves. This was vaguely unnerving, like being out of your depth in water where you can't see the bottom. As we crossed one particularly wide ravine, I asked Chet how much longer the trip would be.

'I don't know, I think my GPS is broken. The pin on the map is right here. Look, Bloukrans Bungy. Wait ... huh? Now it's saying we've gone past it.'

The mystery was solved as we reached the end of the bridge, where a small sign on the side of the road directed prospective bungy jumpers to pull off onto a dirt road and circle back around. The company's headquarters and a few lookout points sat slightly farther inland along the edge of the ravine, giving observers a view of the bridge with the ocean in the background. Jumpers would go from a platform built on the underside of the very bridge we'd just crossed.

Every instinct I had was telling me this was reckless and stupid. If I – if *any* of us – were meant to be descending and exploring the valley, you better believe someone would have built a tarmac road going down one side of the crevasse and up the other.

After parking, we walked over to the nearest viewing platform as a group of jumpers took their turns. Even those who dove gracefully were quickly robbed of all physical agency and rag-dolled through the air, their heads lolling, their hair and limbs alternating between hanging lifelessly, and being jerked erratically to and fro. This was deeply unsettling; I suspect we have some primal intuition telling us that if you see such a phenomenon in the wild, the victim is either very sick or, more likely, dead – and *you're* probably next.

Uncharacteristically, I found myself trying to conjure some positives for balance. There weren't many. All I managed to drum up was the thought of how cool James Bond (played by Pierce Brosnan) had looked when he'd jumped off the dam in *Golden-Eye*. Dad had bought the movie's official video game along with a Nintendo 64 back when I was a baby, but my mom had made him retire it as soon as I was bipedal. By the time my sister and I discovered the console in our basement, it was over a decade old and way out of date, yet we still weren't allowed to play. Naturally, we smuggled it upstairs any time we knew mom would be out of the house for more than an hour or so. I'd spent dozens (more like hundreds) of hours playing the first level, in which you find yourself dropped into a top-secret facility somewhere on the

Arctic Sea. You mercilessly slaughter hundreds of ushanka-hatted Soviets as you make your way through their labyrinthine testing facilities and out to the top of the dam. Naturally, you've brought your portable bungy cord, which you fasten to a railing before gracefully swan diving into the foggy abyss.

The second level, if you're wondering, begins with you crawling through an air duct. You come to a vent and carefully peer down to see an unsuspecting enemy soldier sitting on the toilet, reading a newspaper. You screw the silencer onto your trusty PP7 pistol and eliminate the pathetic shitter, who crumples forward, stupid pants around his stupid ankles. *Hasta la vista, comrade.*

This image, the swan dive, that is, was surprisingly persuasive: it really doesn't get much cooler than that. While it was unlikely I'd get to shoot any commies today, the bungy jump seemed to be the best chance I'd ever have of re-enacting at least a portion of that formative adventure. While I was mulling things over back in the parking lot, Chet weighed in with another one of his zingers.

'Hey, dude, how you feeling?'

'Not great. I kind of feel like I'm gonna puke.'

'Hmm. Well, I have an idea for you.'

'Yeah?'

'How 'bout you grow some fucking balls?'

'Cool. Thanks for the help.'

'No, but seriously, here's what I always tell myself: if a fat grandma can do it, then I don't have any excuse.'

I had no rebuttal. He was right, and it stung. I grabbed my wallet and stormed over to the main office, signing up before I changed my mind. It was decided: I was going to pay a hundred bucks to possibly jump to my death. I want to believe I did it for James Bond, but my best guess is the obvious one – peer pressure. I was *desperate* to not give the [well-endowed] Chet any more ammunition. In any case, I'd finally have an answer to the age-old parental admonishment: 'So, if your friend told you to jump off a bridge, you'd do it?'

'First off,' I'd say, 'Chet's not my friend. But dammit, he has a way with words.'

After paying at the till ('NO REFUNDS!'), I was sent to the weighing station, where my weight was scrawled on the back of my hand in permanent marker as if I was being funnelled through a slaughterhouse. I was fitted for a harness by a gruff, heavyset man, who guided my legs into a sort of sumo belt made from seat-belt fabric. A strap ran from my groin area up through a pulley hooked into the shed's ceiling, while he held the other end. Evidently, 'There mustn't be even half an inch of wiggle room in the crotch area' was the directive he'd been given – he adhered religiously. He yanked his end of the strap downward with heartless venom, stomping on the ledge next to me for leverage. I was wrenched airborne, my feet pulled almost a foot off the ground by the unpitying thong. My balls (the ones Chet had called into question) were rocketed upward in the violent impact and, with no space left to occupy, were forced to retreat fully up into my pelvic cavity, shrunken scrote in tow, leaving me with what felt like a sort of inversion recess. I squealed – at an octave higher than I could have reached seconds before – but I was now fully prepared to jump, new ovaries and all.

A guide marched our group of fifteen or so from the main building out toward the bridge, which was supported by a massive cement archway rooted lower down in the valley. We would be jumping from the top of this arch, which peaked about ten metres (thirty-three feet) below the underside of the road. This meant that my dependence on chicken wire for the day had not ended with the shark cage, as I'd been hoping: to get out to the jump spot, we'd need to walk along a narrow walkway made of aluminium fencing that hung beneath the highway. 'Don't look down!' kept being passed back down the line from up ahead, but it was far too late. After what felt like several hours, the unconvincing walkway gave way to the concrete arch which now converged with us. At least now we were on solid (albeit gently sloping) ground.

Like the dive crew, it was clear the bungy operators had also been trained to raise the spirits of terrified tourists, keeping us distracted with casual conversation before helping us leap to our deaths. Afro-pop music was blasting (cranked up *just* loud enough to drown out any second thoughts). Our names were called in slow succession. At the edge, an assistant would strap the jumper's feet together and connect them to the rubber cable, which, for reference, was about the thickness of a hangman's rope (or, if you prefer, Chet's hog. Allegedly). I'd noticed chunks of this cable in a

basket by the checkout counter back at the main building –
for a few dollars, you could take one home as a souvenir. On
closer inspection, I learned that it was not one solid length
of rubber as I'd assumed, but hundreds of rubber filaments
bound together. The cashier told me that every segment for
sale had actually been used for the bungy; decommissioned
ropes were sliced up and sold to visitors. I was assured these
retirements were *not* dishonourable dischargements, but
merely routine and pre-emptive.

Eventually, my name was called, and I hobbled over. The
harness had cut the blood supply to my legs, but I was too
scared to care. The crew talked to me as they strapped my feet
together, but I was so focused on maintaining bowel control
that I only heard bits and pieces. 'Remember, sir, please dive.
It's very dangerous to jump feet first. And we can't push you, by
the way. People ask us all the time, but we aren't allowed. You'll
have to do this yourself.' Although I'm sure they could see my
knees knocking together (and my buttocks clenched so tightly
in fear that they'd merged to form a singular, seamless glute),
they graciously pretended that I looked confident enough to
get creative: 'Oh, and you can't get a running start, either.' They
probably tell everyone the same thing, but I was still flattered.
Once they had me totally prepared, they encouraged me, repeatedly, to move closer to the edge. But I didn't know where their
stupid edge was, nor did I want to find out. My gaze remained
locked on the horizon – as it had been since we got off the walkway. The old adage of 'Look before you leap!' was out of the

question. I knew that dropping my line of sight, even by a few degrees, would lead to certain surrender. I relished those few remaining inches of concrete – as far as I was concerned, they represented all the time I had left on this earth. I shuffled forward in microscopic increments, trying to delay the inevitable. Eventually, though, the crew's requests ceased, and were replaced with:

'OK, there you go. Three … two … one … jump!'

I complied, launching myself head first off the edge and into the fog, James Bond-style.

'I DID IT! I FUCKING DID IT!' I screamed into the abyss as I fell. My dive was good, at least, so no whiplash for me. But a greater relief was that, despite my terror, I'd not released a cloud of aerosolized shit on the way down, as I'd have been dragged through it repeatedly, like a floating spray-tan. Indeed, I'd clearly played too much GoldenEye; I'd only visualized one smooth, controlled descent, cutting straight to a peaceful arrival at the bottom. But I was nowhere near the finish line. The *first* free fall was over quite quickly, but I hadn't anticipated the second, let alone the third, fourth, fifth, and sixth. The cable pulled me almost back to (what felt like) the height of the bridge, then slightly lower each subsequent time. By the third stomach-emptying descent, the novelty had well and truly worn off. And it turned out the fog wasn't as dense as it had looked from the road. As I plummeted back down each time, I could see nothing but the rocks rushing up at me from below – this image haunted me for weeks.

Once the bouncing stopped, I was left hanging upside down like a slain deer with meat hooks through its ankles. After a few minutes, two of the nimbler crewmen – *Wait a second, they had helmets! Why hadn't I been given a helmet!?* – abseil down to hog-tie my hands up to my feet. This at least allowed the blood to drain back into my body. Now, though, I was forced to wait another eternity while the team above winched the three of us up to the bridge. We made it, eventually.

Before swinging me back over the cement, however, the crew began fiddling with my harness and the surrounding wires. Suspended here, I was now about a hundred metres further from the waiting rocks than I would have been had the rope snapped at full extension. (Not that I'd be more or less dead either way, but I'd rather spend less time at terminal velocity.) Oh, and I'd now be falling *after* everyone had already congratulated me – how humiliating! Worse still, given my ligatures, I'd be unable to control any aspect of my descent, making my literal fall from grace even more pathetic. Rather than going out looking like James Bond, I would plummet instead as a hog-tied submissive awaiting a paddling from his leather daddy:

1995 – 2014

~

He died doing what he loved.

After what felt like another eternity (a feeling that was becoming all too familiar) the crew swung me back over the cement, and I could finally exhale – almost. There remained the small matter of shuffling back along the walkway to solid land. I'd heard that exposure to one's fears was supposed to help conquer them, but this hadn't kicked in yet; the return journey was no less awful than the first. We stumbled to the car, utterly exhausted. I noticed my left eye was bloodshot and its lid was spasming. I figured that I had clenched my face so hard on the way down it felt like I'd ruptured some part of my optic nerve. I hadn't, of course, and recovered in a few hours, although I still get an occasional twitch in that eyelid when I'm tired. A worthwhile trade, I think, for having been reminded, albeit by force, to appreciate the time we're given.

It was a good thing my eyes weren't out of action for long, as we had a safari drive booked for the next day.

RHINOPLASTY

We left our hostel (just a normal one this time, instead of a train) near the bungy bridge at dawn and made good time, arriving early enough to get some lunch and take in our surroundings from the deck of the private game reserve's main lodge.

In the foreground, a herd of antelope grazed over a steppe that spanned a few acres and stretched back for about a mile before sloping steeply upward and giving way to a series of rocky foothills. Looking back at us from a levelled-out plot halfway up the nearest slope was the private lodge of the reserve's owner, who had a full-size helipad in the middle of a manicured front garden.

Parks like this one were common in the region. In most cases, a mysterious foreign investor buys up several thousand acres, then imports as many herds of animals as they can fit in the space. And there's usually a spa and other luxury provisions, too. Their websites encourage you to enjoy a glass of wine while sitting in your jacuzzi or on your suite's deck, where you're perched just above a tableau of the African wild: a watering hole where buffalo and zebra peacefully congregate to bathe and drink. (Or, depending on what you're into, a clearing of dried grass where the local warthogs like to mate.)

We were told that the higher-profile animals on the property, rather than having been bred for captivity, were either rescues or were being rehabilitated for eventual release into the wild. As you drift in and out of sleep in front of the lodge's cozy fireplace, this notion seems charming, but as you load up into the open-sided vehicle and trundle off, the potential downsides become clear. Indeed, had I gotten my hand ripped off the day before, I'm not sure it would have made me feel much better to find out the shark was a rescue: *'He's usually so great with kids!'*

Our guide stopped us just a hundred metres or so from the main lodge and brought our attention to a herd of about a dozen springboks, the small antelopes which are the national animal of South Africa. He told us about *pronking*, their bizarre but endearing straight-legged leaping, whereby they bounce up to four metres (thirteen feet) into the air as they dash across the plain. Other species of bovids – gazelle, deer, sheep, and goats – do a version of this, too, but the springboks have made it something of a trademark. The exact reason for this behaviour is unknown, but it appears they often leap for no reason other than to show off – *pronk* comes from the Afrikaans *to prance*. But that could still be adaptive: as potential prey often use displays of their athleticism and agility to ward off lurking predators by showing they would not be worth the effort to try and catch.

We then made our way over the grassland before rolling to a stop near a rhinoceros that lay resting in the dappled shade of a small thorn tree.

'He's one of the only males of his species in the region,' the guide said. 'He's new, so he's not too friendly with the others yet.'

'The others? So, you've got more rhinos in the park?' I asked.

'Yes.'

'Oh, cool! Do you think we'll see them?'

'Perhaps.'

'How many are there?'

'I'm afraid I can't tell you.'

I was thrown – he'd been so genial just seconds before. But he quickly explained that poachers had been known to infiltrate safari drives such as ours, using them as reconnaissance missions to prepare for illegal trespass and hunting.

'The rhinos get killed just for their horns, you see, so we either dye the keratin of the horns pink, or even just saw the whole thing off. Don't worry, it's painless,' he added, seeing our shock, 'and it's for their own safety. It makes them worthless to the poachers, so they don't bother touching the animal.'

Sure enough, there was only a flat, sanded stump where the rhino's horn had been.

'They're very strong,' the guide continued. 'He could easily flip us over.'

This was easy to believe; the disproportionate bulk of his head was striking. It was the size and shape of a small refrigerator. Driven by his large, stocky frame, it would clearly have made an effective wrecking ball. This was the first time I'd ever seen a rhino in person, and I was captivated. His lack of horn made him no less dignified. His power was undeniable, yet compared to the shark he was slightly tougher to take seriously as his skin, while clearly thick and tough, bunched up endearingly in rolls around his joints, and folded over itself along his back and torso. Such excess puppy fat seemed incongruous for a wild animal who's survived to adulthood and was now scraping for its survival each day, subsisting on dry vegetation. From a short distance away, it reminded me of the ridiculous bundles of skin hanging off a basset hound – although I'm sure I would have felt differently had he decided to charge us.

'We're currently trying to find him a mate,' the guide said, 'but, as I said, he's not feeling very friendly. His family was poached last year.'

The rhino regarded us with his sleepy, doleful eyes. I now became self-conscious about our intrusion, and wished we would move on to leave him in peace. I didn't know how strong his vision was, but it was conceivable he might mistake us as his family's poachers. Who was to say he wasn't already eyeing us up, shrewdly calculating which of us was the weakest link? Evidently, the driver did not share my concerns; because he'd cut the engine and was continuing to rattle off fun facts while smiling cheerfully. So I decided to run the numbers myself. For my money, the rhino's best bet was the gangly, balding man in the row ahead of me. Although he looked to be in decent health, he was travelling with his two young daughters who would, I wagered, be a massive hindrance. Even better, he'd inexplicably chosen to wear rubber flip-flops. This gave me a swell of confidence. You don't have to be faster than the prover-bial bear, after all – you only need to outrun the slowest guy (and/or his kids).

Today though, the flip-flop man would not be held account-able for his choice of footwear: the rhino meandered away, finding a new shrub to munch, and the guide moved us on. We pushed deeper into the park, trundling through steep, denuded foothills dotted with shrubs and wildflowers. On the way, we passed two men on matte-black ATV's. They wore olive-green camo tactical gear and balaclavas, and had scoped rifles hang-ing across their backs. Our guide gave them a thumbs up, which they returned before continuing off into the depths of the park, taking a narrow trail that was clearly for their use only.

'They look scary, but trust me, they are the good guys,' said the guide. The park didn't just protect the rhinos by reliev-ing them of their horns, they took more aggressive measures, too; squads of privately-hired contractors patrolled the reserve around the clock, with license to kill any unauthorized tres-passers. The guide told us that, as a safety measure, none of the reserve's regular employees, himself and the other guides included, were privy to the rangers' identities or whereabouts in the park.

'You have to pay for them?' someone asked. 'Shouldn't the government help? Don't they want to protect their animals?'

The guide chuckled. 'The governments? They are too busy worrying about their own land. But of course they have a different idea about all this. They accept that the poachers are going to kill a certain number of animals no matter what. So they try to at least make some money from this, which they can use to protect the other animals. Think of it as a sort of sacrifice. The big game hunters you hear about – you think they do not have permission from someone in the government? If you bring enough cash, you can poach all you want.'

We came to a sudden halt at the base of a narrow gully, through which ran a snaking, shallow brook. Around ten metres to our right, beneath a low acacia tree near the bank, lay three lionesses. The two smaller ones scattered as we arrived, trotting up over the hill and out of sight. The third – their mother, the guide said – stayed put. She lifted her head up to examine us, but quickly lost interest. Her thick tail swatted half-heartedly at a swarm of flies. The guide explained her despondence.

'She's desperate to mate, so she's a little bit grumpy at the moment.'

On cue, a male lion appeared at the crest of the hill. He posed on the ridgeline, silhouetted against the sky, surveying the contents of the valley. Everyone in the vehicle scrambled over to get a glance of him, fearing he might retreat. Clearly, though, he relished the attention and began descending toward us with a swagger. He hopped over the stream before beelining over to the female. Although she'd ignored his descent and still made no effort to greet him, he'd clearly decided he was the man for the job. Without so much as eye contact (let alone a kiss hello), he mounted her. She didn't bother to reposition from her resting state, but he got to work nonetheless, jackhammering away with pornographic gusto. She gave no indication that she was getting any degree of pleasure, nor that penetration had even been achieved. For a few minutes, he persevered, but it became clear that he was getting disheartened. I couldn't help but feel for him; we've all been there. (*Right, fellas? ... Fellas?*) I tried to catch his eye, hoping to offer a sympathetic smile, but he was too focused on the task at paw.

The guide informed us that these two would be at it several dozen times over the next couple of days while she was in heat. *No wonder they seemed sick of each other.* What I'd have assumed to be a dream gig now sounded pretty brutal. Think of the chafing; they both must be rubbed raw – and all that dust must get *everywhere*. This was a labour of love in the truest sense. Our male was not driven by smouldering lust, nor a desire to deepen the sensual bond he shared with his lifelong mate, but merely by soulless, biological compulsion. Nothing more, nothing less. With hard hat on and lunchbox in hand, he'd show up each morning and punch his timecard, ready for a long, thankless day of back-breaking work while his apathetic partner lay in the dirt beneath him. Nudging Chet, I pointed to his fellow industrious gigolo.

'Interesting. Looks like he goes about his work very quietly – a silent assassin. He's able to get laid without letting the whole park know.'

'Yeah, well, she's clearly hating it. If there was anyone better to do the job, she'd ditch him. Bet you he's got a tiny pecker.'

I regretted starting this proxy war, but my honour was now at stake:

'Nah, she's not bored, she's just dropped the façade of trying to look "sexy". Honestly, I'd be more worried for him if she *was* going crazy. She'd probably be faking it. That's even worse.'

With both our gazes still locked onto the lovers, I tried to gauge his reaction in my periphery. I don't think he bought it. Despite the confidence with which I'd delivered my assessment, I hoped that the cats would offer my theory just a *little* support, by showing us that we were watching tender, passionate sex. This was, after all, an unprecedented chance to thwart Chet on his own turf (i.e. mating). Maybe, just maybe, what we were seeing was just a brief role play with which they liked to kick things off (one or both of them was some sort of necrophiliac). Any second now, she was going to take control and there would follow a sustained period – at least fifteen minutes, ideally – of synchronized, impassioned writhing.

Unfortunately, we got nothing of the sort. It could even be argued that things got quite a bit worse: with her suitor in mid-thrust, the female stood up and shrugged him off before shivering away a layer of dust and wandering up the hill. There was no evidence either of them had reached a satisfying conclusion.

Then, to my great dismay, the wheels *really* came off. The male – hunched over, eyes unfocused, tongue hanging from his drooling mouth – continued his rhythmic pumping even in her absence, as if to theatrically drive home the point that their coitus really had been nothing more than a chemical chain reaction started hours ago. His exertions were nothing but rote, reflexive muscle spasms, like how an octopus tentacle continues to writhe even after being cleaved from its owner.

The safari truck shuddered briefly to life, snapping us out of our voyeurism. The guide had taken his seat and had tried to start the motor.

'We're just running a little bit behind schedule. I didn't realize the time, but we don't want to be stuck this far away from the lodge after dark.'

We'd been so captivated that we hadn't noticed the sun drop below the ridge to our right, casting the valley into darkening shadow. Someone then pointed up to the top of the hill, where the rest of the pride had appeared and were staring down at us. The guide glanced up at them while trying the ignition again.

'During the day, their vision isn't great. They see the vehicle as one large animal, so they would never think of being aggressive toward it. But at *night*, their vision is ten times better than ours, so they're able to identify each of us sitting in the truck and realize that we aren't a threat, we're just a herd of prey that has trespassed onto their turf.'

As he explained the situation, the engine stalled for a second time, and his cheery tour-guide voice cracked.

The lions, numbering seven or eight, had slowly begun to approach. Then it dawned on me: the male's pelvic thrusts, which had seemed so pitiful, had actually been intentional, and sinister. He had a flair for the dramatic and was *taunting* us, foreshadowing what was to come if we were stranded out in his territory. It was the African-bush equivalent of making eye contact with someone across the prison yard and seeing them draw their thumb across their neck. Thankfully, for all our sakes, the vehicle rumbled to life before the lions' vision received its full nocturnal upgrade. The guide reversed us slowly, never taking his eyes off the predators.

 è&

It felt, as it had in the shark cage, as if we were on another ill-advised mission to check whether Darwinism was still in operation. We've made life so sterile and comfortable and monotonous that it's become unsatisfying, so we manufacture scenarios to scratch some primal itch, giving our brains a taste of the stimuli it was engineered to handle. Occasionally, this makes for some memorable incidents, such as when trophy hunters get trampled or mauled by prey that decides to retaliate. But by the time the animal ensures it's the last time that man will ever hunt, the creature's victory is bittersweet (from nature's point of view), as most of its family, and indeed species, will have already been slaughtered. Indeed, despite often putting ourselves in these vulnerable positions, we're spiteful, sore losers. The lion, gorilla, orca, or whatever other captive or wild creature will be swiftly executed if, impelled by its genetic programming, it dares to take the bait.

Animals aren't the only collateral victims of our unquenchable thirst for control, of course. We are, and have always been, far more sadistic than required in the human-versus-human field, too, finding incomprehensibly evil ways to kill, subjugate, and exploit each other. We were reminded of this on the return leg of our road trip, when stumbled into a sort of time capsule.

MIGHT MAKES RIGHT

An hour or so outside of Cape Town, we noticed a farmhouse set back from the road up ahead. 'Breakfast All Day Long!' a small billboard advertised. A papier-mâché strawberry the size of a golf cart had two eyes and 'Fresh Fruit!' painted on it. Scattered along the long gravel driveway was an assortment of farm vehicles and machinery in various states of rust and disrepair, some of which were attended by life-size mannequins dressed in farming gear. There were also human-sized sculptures of other various fruits and vegetables, along with large arrow signs cut from plywood directing our attention toward the house, and mostly empty parking lot. It felt as if we'd stumbled upon one of those funfairs or circuses you find in abandoned Eastern Bloc towns, their rides and stall fronts dilapidated but still standing. Maybe I was delirious from exhaustion, but I wanted to explore; the others were curious, too. *What the hell,* we figured, we had a big enough group to defend ourselves. We parked, and a blonde hostess appeared from inside the building.

'Welcome. If you want to see our farmers' market, it's just to your left. We have lots of fresh produce, as well as some lovely souvenirs. If you're here for the restaurant, please follow me.'

She led us around to the rear of the building and up some stairs. The seating area was an outdoor terrace that overlooked the rolling vineyards and fields behind, which became a mountain range in the middle distance.

'Lovely view, isn't it?' she said, handing us our menus. 'You can see where all our fresh fruit and veg come from!'

As we looked down below, we realized what she actually meant by this: watching the fruit get picked was an actual selling point of the establishment. Between the tilled rows of vegetation beneath us were bonneted black women, each carrying a woven basket they filled as they worked their way down the length of the field. We looked over at the tables around us and saw a few small parties, all of them white South Africans. Though there were a few glasses of water among them, and maybe some juice and a cocktail or two, none of them had even ordered food. In fact, they had shifted their chairs around their tables in order to get a better view of the workers below.

I noticed several of the workers were more lethargic than others – one woman, in particular, was remarkably so. After watching her enjoy several minutes of rest, this started to rub me the wrong way. As a paying customer, I felt affronted. (When in Rome, I guess.) She didn't even respect us enough to even *try* to look busy. I may resent being taken advantage of, but I'm no monster; I soon began to fear for her safety. When I turned to the other tables, however, there were no white-knuckled fists, nor any streams of racial epithets being delivered through gritted teeth. I looked back down at the woman and, upon noticing a handful of straw protruding from her shoulder, realized that 'she' was actually a scarecrow. Scanning more closely, I now saw there were a few others like her scattered between the rows of crops. This made a little more sense, but still, why had they painted her burlap skin black? Who was *that* for? Were the local wildfowl detail-oriented history enthusiasts who required a certain standard of historical accuracy? A Caucasian effigy just wouldn't have rung true?

We scarfed down our food and drove home, bewildered. The labourers, I assume, continued harvesting, paying no mind to their stationary colleague. We had no idea what we'd stumbled into. Was this some sort of role play for locals who were pining for the 'good ol' days'? Some sort of Renaissance Fayre for fans of chattel slavery? Maybe it was a way to let junkies get their fix in a controlled space with legal immunity, like those drug clinics you hear about. This wasn't the craziest theory; it still happened all the time back home in the United States, where sororities and fraternities all across the South were known for hosting Antebellum (i.e. 'before the [Civil] War') parties.

જાજ

Growing up in Lancaster, Pennsylvania, just north of the Mason-Dixon line (which once divided the slave-owning southern states from the free-soil northern states and remains a nominal political-social boundary), meant the infrastructure and legacy of slavery was less evident in my community than in much of the South. We could pretty much brush all that uncomfortable stuff under the carpet, especially because Lancaster had also been home to a number of prominent abolitionists, and a stop on the Underground Railroad, the secret escape route of safe houses and hideouts for slaves travelling north to freedom. Confrontation with our gruesome, and uncomfortably recent, history was, thankfully, not necessary; that work had been done more than a century ago. This allowed for a certain insulation, an ignorance, when it came to the unhealed wounds still affecting so much of the rest of the US – not to mention the mechanisms of oppression that still exist, like gerrymandering and voter suppression. Dwelling too much on the past is unhelpful, to be sure, but until there's been a meaningful, honest reckoning with it, it's difficult to see how reconciliation, let alone true equality, can be reached.

I'd been thinking about all this a lot since I arrived in South Africa – I'd started to learn that, if nothing else, living abroad gives you the time and space to see your *own* country more clearly. On my first weekend I had joined a walking tour of Cape Town's historically significant sites, which gave me a better appreciation of the horrors committed by the apartheid regime. I knew that the black population had been 'relocated', but I hadn't fully internalized what that really meant.

One of the most notable examples was District Six. What had been a thriving, multicultural neighbourhood in downtown Cape Town in the early twentieth century was determined by the apartheid government to be a waste of valuable real estate, and was thus rezoned as land for whites only in 1966, meaning more than sixty thousand locals were evicted, leaving them homeless or forcing them into nearby townships. It was decided that it would be more efficient to bulldoze their homes, shops, churches, and schools rather than to renovate each property. When the time came to actually develop on the newly razed land, however, things went floppy. Today, recovery has begun, but there remains a scattering of grass lots, parking garages, derelict buildings, and thousands of families who will never get their cherished homes and communities back.

The final stop of the tour was ostensibly a more uplifting one, even if it didn't look it at first. We stopped to gather around a chunk of concrete that had been erected at the entrance of a wide, pedestrianized street. It was ugly and bleak, not at all in keeping with the verdant, relaxed aesthetic of the plaza it obstructed. It had some spray-painted graffiti, but not enough, in my opinion, to qualify it as a modern art installation. Rusted ends of thick steel rebar stuck out of the slab's rough edges – like someone had smuggled in concrete debris from Chernobyl.

While perhaps not as thrilling as a giant chunk of radioactive concrete would have been, the artefact still had historic significance: it was a piece of the Berlin Wall that had been presented as a gift to Cape Town soon after the fall of apartheid.

The Germans, having been unified as one country a few years before, were congratulating South Africa for (ostensibly) following suit. By the time our tour group stumbled upon this souvenir, South Africa's 'wall' had been toppled for a quarter of a century – which really wasn't that long ago, when you thought about it.

The apartheid regime was allowed to fester long into the information age for a couple of reasons. For one, the country's geographic isolation made outside intervention costly and difficult. Moreover, other genocides, civil wars, and humanitarian crises across the continent overshadowed South Africa's issues. In comparison, the country had the façade of structural stability: its economy was thriving (probably due to three-quarters of the population providing cheap labour for the remainder), and it had a white government, which was palatable for the traditional powers. For the United States (and its NATO friends), where the Cold War was a far more pressing concern, Soviet endorsement of the South African anti-apartheid effort provided an easy excuse to side with the apartheid regime. In fact, it was the CIA that revealed the location of a young, disruptive Mandela to the apartheid government back in the 1960s. Until the issue's notoriety reached critical mass and it became politically expedient to do so, it wasn't really in the interest of any world power to intervene. Even then, attempts were mostly flaccid.

After the regime change, the international community drew a line under the issue. The values of freedom and democracy had been sufficiently installed, and the country could be left to sort itself out and forge onward as a modern, self-sufficient society. Mandela's inspirational ascent signified a healed nation.

Of course, South Africa had not undergone any true revolution from within. In the end, apartheid was brought to its knees not by Mandela's smile or rousing speeches, by outside pressure and economic sanctions; the machinations were far more politically and morally complex. Not that the West cared – we had our fairy-tale ending.

Those who actually had to live in the country were able to enjoy no such illusions. The man I met on the plane was right; the honeymoon period was over and people were fed up. There were a few wide-scale protests that garnered media attention during my visit, during which the local highways were closed off by police or blockaded by piles of burning tyres and groups of marchers. The government-issue euphemism for these incidents was 'Service Delivery Protests'. It turned out that these 'services' referred to basic human rights like water and healthcare, making the citizens' anger totally understandable, in the view of us volunteers. Still, we were mostly insulated; Muizenberg, whose inhabitants were mostly white and middle class, was not deprived of basic infrastructure like the nearby townships were.

֍

Unfortunately, with most of my time spent between the school children and the other volunteers, I didn't have the

chance to canvas many locals of voting age. The adults who
were in my orbit had no time to spare: Charles was (allegedly)
busy enacting vigilante justice, Pastor Samuels was trying to
stick to his rigorous siesta regime, and the Shaggy impersona-
tor was deflowering as many of his groupies as possible.

Still, I did my best to put my finger on the pulse once
in a while. This proved easiest in taxis, whose drivers, almost
exclusively black, were usually politically informed, and always
receptive to an informal interview. They were so talkative, in
fact, that it was *me* who was seen as a wet blanket. For the first
several rides, I acted as I would have back in the US: the driver
pulls up, asks your name, and that's the last of the conversation until you thank him as you get out of the vehicle at your
destination. This code of silence was customary, and often
mutually beneficial. Many drivers seemed happy to sing along
to their music or take phone calls with friends or family. Here
in Cape Town, though, I noticed that I was accruing substand-
ard app reviews. *But why?* I hadn't drunkenly vomited on any
of these rides, nor had I cancelled or stood any of them up.
I then tried dialling up the geniality, starting these rides off
with some chit-chat before lobbing out a few softball questions to get them going on current affairs or sport. *Like the
sword from the stone.* The floodgates opened, and I was inun-
dated with life stories, local gossip, and political insight.

A worry for many of them was xenophobic violence –
many were migrants themselves from other parts of Africa,
and populist rhetoric had stoked the fires of distrust for

outsiders. And this problem wasn't going away. For all its troubles, South Africa is a veritable paradise compared to much of the poverty-stricken and war-torn continent (and this imbalance will only grow as climate change makes more of the continent inhospitable). As such, Cape Town and Johannesburg will likely balloon in population over the coming decades, adding stress to an already creaking infrastructure.

My friend from the plane was also correct in that, while there was little optimism regarding the government, it retained a staunch loyalty nonetheless.

'Yes, they're all corrupt,' one driver said, 'but so is the rest of the world. It's every man for himself.' I remarked that he didn't sound too troubled by all this, and that, regardless, the instances of brazen malfeasance must have been frustrating. I noted that one of the ANC's leaders had been arrested earlier that week after having been caught embezzling millions of taxpayer dollars over the last decade.

'Well, I'm just happy a few of us can be rich,' the driver explained. 'We were nothing for so many years. And now we're in power, living in big houses, driving nice cars.'

'We' sure seemed like it was doing a lot of heavy lifting there, but his feelings were by no means unique. For much of the black population, years of dehumanization, with their lives being reduced to a binary, life-and-death struggle had eliminated nuance from their political outlook. Given a choice, many would prefer a 'bad' black person in power to *any* white person, if only because this signified a power shift, however marginal, away from the white hegemony. This sort of belief only prolonged the life of their broken system, of course, but even this Pyrrhic victory was irresistible.

Predictably, rabid tribalism is liable to be taken advantage of by opportunists who have no emotional connection to (and are even disdainful of) the poor and working class. For example, the 'white monopoly capital' argument, a term championed by Jacob Zuma, and, more specifically, by his almost entirely white public relations team, was used to deflect criticism by insinuating that the country's *real* enemy was not its elected officials, but a cabal of white South Africans who still hoarded an outsized share of the country's capital. By playing this inflammatory card, Zuma (and the Guptas) bought themselves some more time in power.

While they and their ANC cadres got rich, the rest of the country was being run into the ground. It was years before any official criminal investigation would be undertaken. By the time the bubble was burst billions of taxpayer and consumer dollars had been misappropriated, never to be seen again. The judgement came in the form of the Zondo commission, assembled to discover the extent of government corruption during Zuma's tenure. Power, transportation, telecommunications, internal revenue, law enforcement, defence – no piece of the pie was left untouched by the fingers of avarice. Even South African Airways, the operators of my first flight into the country, had been implicated, and would eventually declare bankruptcy.

இ

As far as *white* locals went, it appeared that they, like the volunteers, were also insulated from the hardships faced by so much of the population. Less than one percent of the country's white population lived in townships. The wealth gap was such that even the *poorest* quarter of the white population was still markedly more comfortable and secure than their black counterparts. For example, even the scheduled power cuts were avoidable for people who could afford their own private generators, which cost a couple of hundred dollars. But I didn't get to spend much time with these locals – I didn't want to hang out with the drug-dealing Poseidon, or go to karaoke night, and the only white taxi driver I'd met was Lollipop. Most of what I learned was collected passively. Unfortunately, I squandered one of the better opportunities for conversation I got, and in memorable fashion, too.

I'd realized that I, like the girl from the Lollipop drive, was in need of a haircut. I did a brief search to find the cheapest barber nearby that seemed legitimate, and more importantly, not candy-themed. (I wasn't interested in needing to steel myself for a visit to, say, 'Mr Candy Cane's Grooming and Waxing Warehouse'.) After finding a place with a trustworthy-looking website, I called over and made an appointment for the following afternoon.

The next day, our pickup from school was delayed and we arrived home late. Luckily, Chet and I had bought a second-hand bike from the local pawn shop, so I still had a chance to make the appointment, but it meant a frantic velodrome sprint along the main road on what was a particularly warm day. I arrived just in time and sat, flustered but relieved, beneath the thick, stifling plastic cape, which the barber had flung around my neck with vigorous panache. As my heart thumped relentlessly, I started to notice the room's stuffiness. The ride had been so intense that my body hadn't had the chance to react – until now.

Sitting before the mirror, I had the rare opportunity to watch myself suffocate in real time. My neck was being completely constricted by the unyielding strip of Velcro. I tried straining to loosen its chokehold, but it was too secure – I risked a hernia if I pushed harder. Folds of skin, given nowhere else to go, now hung, bulging, over the collar, giving me a sort of turkey's wattle. I hung in there for as long as I could, panting and spluttering to the barber how I'd like to 'keep a bit of length on top, if that's all right', hoping the wave of discomfort would pass. Alas, the greenhouse effect created under the tarp had triggered some sort of emergency protocol; my body activated sweat glands that hadn't ever been called into action before (and haven't been since). The beads of sweat that had begun to appear when I entered the building had now given way to rivulets that pissed down my beet-red face. My vision started to blur and constrict – I couldn't last another second.

'Excu— Sorr— Sorry, I gotta use the bathroom.' I sprang up, ripping off the collar. The black vinyl of the seat was drenched with sweat. I sprinted to the sink and ran my head under a cold tap for a couple of minutes. Keeping my head beneath the water, I reached for the paper-towel dispenser, fumbling with it blindly before realizing it was empty. I squeezed as much of the water out of my hair as I could before slinking sheepishly back to the shop floor. Both barbers were visibly unimpressed, and I couldn't blame them. Alas, because I looked so *very* guilty, I faced the same dilemma as the Lollipop man: offering *any* explanation would only make things worse. So I stayed silent as I sat there awkwardly, my mousy hair matted down my forehead like a sodden, red-faced Hitler (sans moustache). I decided to pay double the fee to make things right, but when I stood up I realized the cash I'd brought must have fallen out of my pocket during the bike ride. I promised I'd be back later on. The barber, sweeping up the wet hair, waved me away without looking up.

'Don't bother.'

I knew I had no choice but to like the cut, as I definitely wasn't going to risk getting another one before I left. It was somewhere between Hitler Youth and Amish bowl cut – by this I mean it was divided roughly half and half. *Oh, well.* I'd just have to make sure to stand sideways if anyone whipped out a camera.

QUIET, PLEASE

After this debacle, I resolved to build up my fitness. Thankfully, I'd packed my tennis rackets. I called the nearest club and learned that they had courts reserved for non-members to use on Tuesday and Thursday afternoons. One Tuesday, school was cancelled due to a particularly well-attended and high-spirited service delivery protest that had blocked all the access roads. But that issue was closer to the city proper to the north; heading south was fair game. I called the club, who said they'd be open as usual. The courts were a few miles away, but I certainly wasn't going to call the Lollipop cab, so I decided to take my chances on the train. It was rush hour, so the carriages were crammed. I felt terribly conspicuous; there was nobody else with a tennis racket, let alone any other white teenagers.

What didn't help matters were the targeted ads (well, what *felt* like targeted ads, at least) like the ones I'd seen at the stadium. Covering almost the entire train's interior, these square stickers were advertising all sorts of services and medical procedures:

- ABORTION (appended in some – but not all – cases, with: SAME DAY, PAIN FREE)
- VAGINAL TIGHTENING

- BODY CLEANING
- BREAST/HIP ENLARGEMENT

(Most ads offered multiple or all of these – that's one hell of a spa day!)

You also had:

- PENIS ENLARGEMENT (appended, in some cases, with: EXTRA POWER, and/or BIG EJACULATION. The duration/specifics of treatment were unclear, and no mention, worryingly, of it being pain free.)
- BRING BACK LOST LOVER
- ASTROLOGY
- FINANCIAL PROBLEMS
- BAD LUCK (I assumed, as far as the latter two conditions went, that you'd be helped *out* of them, rather than be *given* them.)

Each of the stickers also had a phone number, and usually a name: DR Z, DR JOHN, DR ROSIE, MAMA WINNIE, PROPHET MZUMBU.

When you mix poverty and a lack of education into a society that already values traditional medicine and superstition, and sprinkle in a dose of distrust for Western medicine borne from years of colonial oppression, you get a population that's susceptible to be taken advantage of by witch doctors (called *sangomas* in South Africa), traditional healers, and other unlicensed practitioners, purveyors of cures for any somatic or metaphysical ailment under the sun. Poor fools. Folks back home would *never* get tricked into giving away our hard-earned money so easily, letting cynical healthcare and cosmetic companies prey on our insecurities and getting us to fork over thousands of dollars for bogus surgeries and spiritual guidance.

Despite the fact that I was interested in at least half of these ads' offerings, I tried not to stare at any of them for too long, lest it appear that I was trying to commit the numbers to memory. That said, a part of me did wonder if it would be better to take this issue by the horns and glare at the ones referencing penis/ejaculate enhancement while frowning and shaking my head, making it clear that I had no need for that whatsoever.

Thankfully, the carriage's other occupants had become distracted. Two teenage boys had forced open one of the sliding doors down at the other end and were taking turns hanging outside the carriage as we hurtled along the tracks. The woman next to me looked at them with weary disdain before turning to her friend.

'They must stop doing that. Last week one of them fell out.'

But I couldn't blame the kids for getting a better view – it was spectacular. On our right, the mountains rose so steeply that their peaks were invisible from inside the train. The slopes descended almost straight into the water, inter-rupted only by the thin strip of paved land at their base along which ran the train tracks. We were close enough to sea level that spray from the crashing waves speckled the port side's windows. Our destination, Fish Hoek (roughly translated to *fisher's corner*), was a few miles south, set into a cove along the eastern side of the peninsula. The town and its pristine beach panned out on our left as the tracks turned, hugging the inward curve of the coastline.

When I arrived at the station a few minutes later, there was no tennis club in sight, so I asked an attendant, who told me that the courts were set back from the ocean in a valley just over some low hills. 'Walk that way for about ten minutes. You'll find them.' Sure enough, I made it, but when I arrived, the club manager told me I'd have to wait in the clubhouse for a little while.

'We can't play right now, sorry.' He pointed to a hill off to the south. An undulating line of glowing orange crawled slowly down the slope, leaving blackened earth and scorched flora behind it. The wind, which was carrying away much of the smoke, had carried the blaze from farther down the peninsula.

'Don't worry,' he continued, 'this happens every so often. The drought's gone on longer than we expected this year. We've called the fire department, and they're sending a helicopter to make sure it doesn't get too close. We can play once they give us the all-clear. You'll fill in for Ted. He's on holiday this week, staying at his game lodge in Zim[babwe].'

The fire brigade soon arrived to slow the advance of the flames and seemingly got things under control. The club members emerged from the clubhouse, divided into small groups, and scattered to the different courts before starting to warm up. I was the youngest there by a couple of decades, most of the other players were sweater-vested retirees. The fire was a minor inconvenience, and the service delivery protest wasn't one at all; neither were mentioned while I was there. This was a Tuesday, and we play tennis on Tuesdays, the way it's always been. Do whatever you want, but don't bother us on Tennis Tuesdays. If you *must* burn, burn in a different direction, for God's sake. But weather conditions around the world keep getting more extreme, and South Africa struggles with drought at the best of times. The flash floods, fires, and famines will spread more easily each year, each time encroaching further into the periphery of Tennis Tuesday. Perhaps one day soon something would have to give, but not today.

The session finished around sunset, meaning I had to take the train *and* walk back across town in the dark. This wasn't much fun, but I braved the journey again the week afterward. The outward journey went smoothly – sort of. I got detained on the way there by two very stern, very portly security officers as I tried to leave the station at Fish Hoek. The man and woman hauled me to a dinky little interrogation room whose tiny, lone window up toward the low ceiling was screened by chicken wire. I'd failed to buy the right type of ticket, they said, and I now faced having to pay the requisite fine or be charged with a criminal offence. They both wore combat boots and high-vis orange vests over khaki army jumpsuits. The man, coloured, grey-haired, moustachioed, paced the room. His female colleague, boasting only fractionally less facial hair, blocked the door while slapping a wooden ruler against her palm. After letting the tension build, the man squared up to me and announced my punishment: a fine, the equivalent of five dollars. If I knew what was good for me, I'd cough up and skedaddle. I obliged.

On the way back, though, the train got held up for over an hour because someone had 'fallen' onto the tracks up ahead. I didn't see the body, but that was enough of an omen for me – tennis could wait until I was back in the United States. Unfortunately, this meant I didn't get to know any of the other players too well, but the candid assessments I heard regarding the country's state of affairs from the other suburbanites I ran into were more than enlightening enough.

A few tennis tournaments were televised during my time in South Africa, and one of the only local places the matches were screened was a pub in the gated neighbourhood on a man-made marina set back a mile or so from the beach. Houses surrounded the edges of the main reservoir and lined the banks of the winding canals. Fortunately, the security guard deemed my racial profile acceptable, as did the private neighbourhood's residents, who concluded I ought to be considered an *ex-pat* rather than an *immigrant* (there might not be much *semantic* difference here, but their stigmas are vastly different). This hospitality did not extend to any of the black pedestrians or drivers who sought to gain entrance – the guard (also black) was often in the middle of interrogating them as I strolled past.

My high-level security clearance also meant I didn't raise the hackles of the pub's regulars. Though I came for the tennis matches, nothing I saw on-screen was as fascinating as some of the conversation I heard behind me. I listened intently, but made sure to never turn around. While certain talking points on race science and its adjacent topics were predictable, what was curious was the tone in which these theories were delivered. The people seemed matter-of-fact and measured, the way you'd talk about the differences between two brands of toasters or make small talk about the weather. On more than one occasion the conversation drifted toward phrenology – I often felt we were just seconds away from someone pulling out a skull and a pair of old calipers.

This may sound hyperbolic but, as I'd learned on that walking tour, the basis for discrimination during apartheid wasn't far off from this. Given the country's wide range of racial and ethnic demographics, it wasn't always clear to the government where certain civilians ought to be placed on the hierarchy of subjugation. For anyone whose race *wasn't* imme diately apparent, the government came up with the 'pencil test', whereby an official would stick a pencil through the subject's hair and instruct them to shake their head around. If the pencil stayed in, you were adjudged to have the hair of a black person and would therefore be classified as such. Anyone who didn't look white but still 'passed' the test was thus deemed coloured (or Indian).

MORTIFICATION
OF THE FLESH
(OR: HOW I LEARNED TO STOP WORRYING
AND LOVE GETTING MUGGED)

I'm not sure what the tipping point was – our pit-stop at the slave farm probably had something to do with it – but I had to figure out what the hell was going on. Why was nobody talking about this? Were they oblivious? Or in denial?

The sheer proximity of the suffering to even the wealthiest, most luxurious neighbourhoods makes it impossible to claim ignorance. Indeed, the hillside wineries on the eastern side of the peninsula, in Constantia, for example, look *directly* out across the Cape Flats and its townships; Khayelitsha, Gugulethu, Langa. Enjoying the view requires looking out over at least one million people for whom going to a wine tasting would be an unimaginable luxury. A million souls trapped in the soul-sucking doldrums, with access to neither the resources of the mainland nor the glamour of the Cape. *This* was their Africa: no safaris, no shark dives, no chilled Chardonnay. For people like me, it seemed to require the same sort of self-discipline as the bungy jumping: Whatever you do, don't look down. Lock your eyes on the horizon at all costs.

Let your gaze drop and you're fucked. Whatever you think you saw was wrong; there's nothing down there – nothing important at least. The screams, hopelessness, the desperate pleas for food and water are a figment of your imagination … *Now, sir, if you please, edge forward. That's right, easy does it.*

So, what? Was that it? It couldn't be that every single person *not* spending their every waking moment talking and thinking about these horrors was a self-centred asshole. There had to be some explanation for our woeful paralysis in the face of these issues. Something biological, surely.

We do, it's true, have an innate aversion to stress and responsibility; always preferring the path of least resistance. We don't want to stand out or rock the boat. Since we're conditioned to protect our scarce resources and our own kin, we find it easier to distance ourselves from anybody we can rationalize to be from a separate 'clan'. As such, the likelihood that we'll intervene to help someone in need decreases as the particular vulnerable group increases in size (which explains why those UNICEF commercials linger so excruciatingly on just the lone malnourished child staring longingly into the camera). In fact, we become less likely to act if even just a *second* person appears, be it another victim or a bystander like us, as we presume (consciously or not) that the chance some- one else will step in and take responsibility grows, too. As a bonus, guilt is therefore diluted in the event of inaction and tragedy. We're also not great when it comes to big numbers – 'a million deaths is a statistic' and all. As we attempt to fathom these distressing global issues, psychological 'tools' (and blind spots) like these allow us to zero our moral scales. South Africa presents a real-time case study, or indeed a stress test of this, and it sure seemed like most people were passing with flying colours.

On that note, I began to wonder whether the land's unique natural beauty was, at some level, disadvantageous. The country offers countless focal points unconnected to indigence, disease, and suffering toward which we can focus our attention and our cameras. Harrowing expanses of helplessness and blight are not only juxtaposed against – but in many cases are *overshadowed* by – colourful, photogenic, *unspoiled* panoramas. Because of the enchanting tableau, a visitor's well-meaning idealism can burnish their impression of the area and dull their concerns; it's easy to imagine that such a paradise could ease *any* suffering. For us voyeurs, feelings of shame can therefore be lessened by distraction, which feels more excusable than conscious evasion. Rather than averting our eyes, we can instead *divert* them. This distinction may seem like semantics, but it leads to drastically preferable moral consequences. It makes it easier to suppress the creeping suspicion that by coming here and doing nothing we are guilty of some undefined, yet undeniable, dereliction of moral duty. And so, the sweeping destitution becomes just another feature to regard, to explore, and to reflect upon – or not.

As if reckoning with all these issues wasn't challenging enough, the constant call for vigilance added another layer of complexity. The fact remained that crime was rampant, something we were made aware of early and often by programme administrators, locals, and the longer-tenured volunteers, all of whom offered guidance on every aspect of our day-to-day lives. This guidance followed the same basic template. One of the first examples came from the guide from our walking tour that first weekend, before he set us loose to explore the city's small plaza, Greenmarket Square. Originally used as a slave market by the Dutch East India Company and other colonists in the seventeenth and eighteenth centuries, the area was now home to a flea market filled with rows of little stalls where locals peddled their (non-human) wares. The guide got the group's attention, declaring:

'Don't explore alone. This area is full of pickpockets. If you *need* to buy something, use the ATM over there. But make *absolutely sure* no strangers are behind you, and that you use it as a group. Oh, and jiggle the card receiver first to make sure someone hasn't put a fake reader in it to steal your card details – that's been happening quite a bit recently. We'll meet back here in twenty minutes. Have fun!'

It turned out what I'd really needed from him was a *caveat emptor*. The first stall I visited, I fell for a matronly African woman's sob story about her escape from her home country. Fortunately, she claimed, she'd brought some fabrics and supplies with her and had used them to hand make each of the patterned crafts before me, with the help of her daughters. Happy to put money toward a worthy cause, I bought a tea towel, nobly refraining from haggling down the price (something I'd been told happened quite a bit on this continent). Seconds after my purchase, I walked past three other stands with similar-looking plump women reciting identical stories to different tourists while peddling the exact same handbags, towels, and beaded jewellery. *Whoops.*

But the guide's briefing was by no means the first of its kind. Within your first few days, you're instructed to: *never* leave your house after dark, *never* carry more than ten dollars in cash on your person, *never* use, or even show, your phone in public in case it gets snatched, and to *always* roll your car windows up before reaching a stop sign or red light (Endearingly, traffic lights are called *robots*, even on official road signage.) in case there are carjackers waiting to pounce. You're reminded to walk in groups as much as possible, and to never make eye contact with, let alone *talk* to anyone who approaches you on the street. If you leave your wallet or purse on a park bench or at the beach, you're scolded for your naïveté, and told that it's not even worth your energy to go back to check if it's still there. (Spoiler: it won't be.)

It was impossible to dismiss these warnings as fearmongering, for along with each 'commandment' came a personal or nearly personal anecdote (this had started, of course, with the break-in and stabbing at the volunteer house). Anyone who'd lived in the country for more than a few months seemed to have a worryingly extensive catalogue of instances where they, their friends, or their associates had been a victim of some type of crime. For the most part, these incidents were petty, but a good number were more traumatic. It was not uncommon for volunteers to cut their trip short, for example, the young man who was stabbed or the woman who'd been mugged, including some who *weren't* direct victims but who were wrung out by the anxiety and pressure of it all. The national statistics seemed to corroborate these anecdotes: there were almost a hundred murders a day, and some of the highest rates of violence, theft, carjacking, sexual assault in the world. And the data didn't even tell the whole story, as many crimes went unreported, due to distrust or simply lack of faith in the underfunded, and, in many cases, underqualified, police force, which was only able to prosecute a fraction of any criminals it *was* able to apprehend.

ðŸ‚

Some tourists and volunteers, however, managed to convince themselves they'd be spared from danger if their intentions were sufficiently pure. This included a girl who arrived a week shortly after me to help at an orphanage in the heart of a nearby township. Noble work by any standard, but she soon decided this would not be sufficient. One night, after she'd been in the country for no more than a week and a half, the programme administrators caught her sneaking out to the highway to hail a taxi that was headed into the nearby township. After wrangling her back inside, the administration team asked her to explain what the hell she'd been thinking.

It turned out she'd promised several of the kids she would have a sleepover with them at their houses to get an appreciation of their 'lived experience'.

'These kids have changed my life,' she explained, 'so I want to get a window into theirs. We have a special connection. I feel like they're my *own* family.'

Her intentions were benevolent, I'm sure, and she should perhaps be applauded for committing so devotedly to her role, but allowing her to pursue her dreams was out of the question. These townships were dangerous for the average resident, let alone a white, red-headed girl from New Hampshire.

Crime being conducted on criteria *other* than the victim's moral purity is a notion you'd think the volunteers would have been disabused of upon hearing that one of them had been stabbed or robbed. If our intentions counted for *anything*, we and our houses would have been off limits to the local gangs. Hell, we might even have deserved *protection* in our dealings around town, sort of like Michael Jackson when he went into the Brazilian favelas to film a music video. But of course we were awarded no leniency and were taken for the easy marks we were.

Bizarrely, though, these experiences were almost welcomed – *relished*, even – by some who, evidently, had a fear far more deeply rooted than that of getting robbed: that of appearing *prejudiced*. With so much of the country's troubled past still visible, many aspiring do-gooders feel compelled to distance themselves – as conspicuously as possible. It was understandable, given the fraught, polarized discourse around topics like colonialism back home in the United States and the United Kingdom. The trouble is, our history lessons only cover the past two or three centuries; moreover, we're only really taught about our own history, which doesn't cast us in a flattering light. At any rate, there's only one appropriate response to this: anyone with a heart and a conscience must show contrition. If you want to identify as anything to the left of 'fascist', you can't exactly proclaim that you're not sorry or that you don't feel ashamed. We therefore go to great lengths to establish that we are good people. And these lengths, we surmise, include: self-flagellation, grovelling, and general fussing and hand-wringing. Whatever you do, the main thing is to look busy; you don't want to be seen resting on your laurels. But without something concrete, some consensus to work toward, people are left to their own devices to figure out how best to achieve reconciliation – or at least to assuage their conscience. It's no surprise then, that they head off in some peculiar directions. In this case, it seemed that gnawing guilt and the thirst for forgiveness had incited a sort of masochism.

It was common practice for the volunteers to hire a taxi van on the weekends to attend a public barbecue in the middle of one of the nearby townships. Although someone would have their wallet, or bag, or phone stolen almost every time, this was seen as almost a badge of honour, and dismissed as just being part of the 'true South African experience', as if being happy for the pickpocket who took your phone would earn you some sort of moral credit with the thief (or his race).

'Losing my phone is nothing compared to what his ancestors were put through.' It was always intriguing to hear the victim's over-eagerness to forgive or downplay the crime, implying that the muggers or burglars didn't know better or couldn't help themselves. Invariably, the volunteers would blame themselves: 'It's no wonder my car got ransacked, I left it unlocked overnight – what was I expecting?' I wondered whether they'd be as proud to recount these incidents if they'd been wronged by a *white* person, or if this had happened back home. (Doubtful; the whole premise of the privatized penal system in the US was to punish people for their socio-economic background.) At any rate, in this context (i.e. 'This is Africa'), it appeared victim-blaming was completely acceptable.

I noticed a similar phenomenon in the region's suburban, affluent areas. Here, private security vehicles patrol wide streets, and otherwise dignified family houses are bordered by austere, three-metre prison-yard walls, whose concrete slabs or cinder blocks are topped with electric fencing, barbed wire, or even just large shards of broken glass that span the length of these walls, silhouetting some sort of bleak, depressing stegosaurus. To many do-gooders, these conspicuous, draconian precautions are a political statement in their own right: they're inflammatory, perpetuating the climate of paranoia at the societal level. Intentionally or not, correlation and causation get conflated here and, given the country's stark racial segregation, this all becomes a simple, damning equation: the more security and protection a residential (read: white) neighbourhood has, the less it thinks of 'outsiders'. For those desperate to show solidarity, then, the contrapositive thus becomes the only acceptable conclusion: a *true* Not-Racist must renounce vigilance and security measures.

For the rest of us, maintaining a constant level of alertness felt like a necessity. After a spate of muggings, you avoid the relevant areas where possible, and you certainly don't walk there alone, even during the day. If your neighbour or colleague tells you there's been an uptick in carjackings on certain stretches of the nearby highway, you begin to eye every roadside construction crew with distrust. At every intersection, you check your mirrors to see if anyone approaches the vehicle. And so on down the list of commandments.

Even just *parking* was nerve-wracking: as soon as you find a spot, you're converged upon by *car guards*, a suspiciously self-justifying phenomenon found anywhere people tended to park – public lots, or any roads near beaches, markets, or shopping centres are hotspots, along with pretty much anywhere else in the city. Individuals, or small, organized crews patrol the pavements and loiter at street corners, waiting for a driver to take an available space before scurrying over to offer their services: they'll 'protect' your car while you're away, and some of them even offer to 'wash' it, too. There isn't a set price; whatever coins you happen to have will suffice. In truth, you don't *have to* give them anything, but as the group of scruffy men swarm around you and your car, it's difficult to feel like you have much negotiating power, especially when you notice one of them is holding behind his back what looks like a crowbar.

'Hey pal, you look like you're going to be in that restaurant for a couple of hours. It would be a real shame if anyone came by and tried to forcibly gain entry into your nice new ride. What's this? The rear window already has a crack in it – somebody could easily break through and open the door from the inside. Hmmm. Anyway, don't worry about any of that. Go enjoy your afternoon, and leave her with us, she'll be in good hands.'

Sometimes, they won't charge you until after the fact. On one occasion, I came back to my rented car to see a man waiting beside it. He informed me that, by parking on his turf, I had implicitly agreed to his terms of service. He'd washed my car, he said, and I now owed him. He clearly hadn't done anything of the sort, but, as I could see a few of his associates beginning to encircle us, I agreed to pay. But my cash was in the glove compartment, I explained, so I'd need to get inside first. Once in, I quickly locked the doors; several of the men yanked snatched at the handles. One of

them tried to yank my windshield wiper up so that it would snap back down and shatter the glass, but he lost his grip as I stomped the gas pedal and lurched away. I sped off to safety, wiper erect, thanking God my parents had made me learn stick-shift (a skill that wasn't really necessary back in the US, where almost all cars were automatic) a stalled engine would have been catastrophic.

Sadly, given the lack of police presence, there's often no option but to accept the terms of these car guards' little game, wherein they've ingeniously created both the supply *and* the demand. So, with sincere thanks, you offer these kind fellows the equivalent of a dime or a quarter before scurrying off, but not before double- and triple-checking that you've locked the doors and rolled up the windows. To enjoy the rest of your day, you must pin your hopes on the notion that these men are operating in good faith, and try to dispel from your mind the realization that, in this contrived 'market' whereby your property is auctioned back to you for pennies on the dollar, an enterprising third party could easily outbid you in your absence. I saw one German tourist pre-empt this, raising the starting bid by handing a crisp fifty-euro note to a car guard and shaking his hand nervously before heading into a grocery store. The other guards around the parking lot were flabbergasted, none more than the fellow who'd just made fifty euros: he hadn't even said a word to the man before he'd passed him the money and hurried away. Not only that, but it turned out he wasn't even a car guard; he was just a random black guy who happened to be walking through the parking lot. I grimaced, as just a couple of clueless high rollers like the German ensured these guards harangued the rest of us mercilessly.

᠔

In fairness to the German, his donation was far from the only instance of misguided philanthropy I witnessed. It felt like there were very few ways to offer meaningful support at an individual level, so people had to get creative. One approach I saw often was to tip black waiting staff a few percent on the dinner bill. But no more than 10 percent; you don't want to come off as patronizing. (The trick here is to smile warmly and give them the coins by hand, one by one – that way they know it's Reparation Money.)

As far as more earnest options went, one was a boat trip out to Robben Island. Despite appearing to be little more than a sandbar protruding from the Atlantic, this little island a few kilometres off the coast of the peninsula had quite a sinister history: it was apartheid South Africa's version of the US's Alcatraz, imprisoning political dissidents for decades – Nelson Mandela was kept there for nearly twenty years. Now a museum and protected heritage site, tourists can visit via ferry to be shown around by tour guides, most of whom are ex-prisoners.

Another popular activity was the Township Tour. This one I could never really get my head around. Given the mysterious, remote nature of Robben Island, it seemed reasonable to me that visitors would want to explore it, in the same way people tour concentration camps from World War II. There's an element of macabre curiosity, for sure, but it's also important for posterity to not let these historical wrongs be forgotten.

But if, hypothetically, prisoners were still being kept there, paying a visit to the island would become a different proposition entirely. Unless you were going there with specific journalistic, or revolutionary intent, it would be strange for anyone (in self-proclaimed opposition of the regime, at least) to sign up for a chaperoned tour through the prison's bleak hallways so as to retch at the smell, or to gawk at the festering mould:

'Wow, that gruel they're feeding y'all really is as lumpy as the brochure described. D'you think it's the damp air in here that makes it congeal like that? Sorry, hold that thought – our ferry's leaving in fifteen minutes, and I was hoping to swing through the gift shop before leaving. Can we rain check? Actually, I probably won't be back anytime soon ... unless they arrest me for those parking tickets – ha! Anyway, try to make the most of the rest of your sentence. Ten years? That's nothing. Oh, you said "life sentences" not "years"? Hmm.... . Oh no – please stop crying. Chin up. Hey, before I go, can I take a picture of me giving you a hug? Thanks. Hey. Hey! Let me go, you mongrel! Guards! He's choking me through the bars! Guards! Help!'

You get the picture. Tickets for these tours were sold by many of the local hostels and tour companies, who were proud to advertise that you would be shown around by a real, live resident of the township. I wasn't quite sure of the intended purpose of these sightseeing tours, but my hunch was they were overseen by travel companies with similar ownership models as our volunteering organisation, preying on Westerners' addiction to poverty porn.

But maybe I'm being unfair; I'm sure the industry creates a few jobs, and the proceeds do help certain locals. And there probably aren't many more legitimate ways for a credulous tourist to interact with and give directly to a resident of these areas. Given the chasm between classes and the demographics that still exist, these tours, providing sanctioned, secure access, are the only way for outsiders to get any exposure to, or appreciation of, the harrowing realities faced by so many in the global South.

Even so, I was tickled by the thought of a gaggle of nervous tourists waiting just outside a slum as matching neon shirts are passed out to them by their guide, who connects them to his dog-sled leash so that they won't wander off. As they proceed, he walks backward in front of them while waving a little golf flag for visibility. After half an hour or so, the tour ends: now's their chance. Elbowing each other out of the way to get to their guide, they excitedly fumble through their purses and pockets to retrieve handfuls of banknotes, which they desperately stuff into his palms.

MONKEY BUSINESS

As much as I dreaded parking in the city, there was somehow a far more intimidating car guard situation in the wild. There were dozens of lookout spots along the length of the peninsula, all of which were accessed by a handful of winding, one-and-a-half-lane roads which, despite doing their best to negotiate the thin, mountainous headland, were often forced to stray precariously close to the cliffs. Where space for roadside shoulders *had* been carved out, warning signs reminded visitors of the treacherous drops below.

But there were other warning signs, too, for a more *active* threat: the troops of mischievous baboons that patrolled the region. Sure enough, tourists returning to their vehicles were often confronted with the dilemma of a baboon blocking access to the driver's door, or even sitting on top of the car. If there were no national park employees nearby, it was up to you to haggle with your fanged adversary.

In one incident, I witnessed a man (who seemed well-versed in these dealings) whip out a packet of chips from his backpack and scatter them on the other side of the car for the baboon to chase, which gave him the opportunity to quickly get inside.

On a different occasion, I saw a woman stumble upon her ape-adorned Volkswagen woefully unprepared. By the time she understood the rules of engagement, the result had already been decided: she'd been soundly beaten. The primate held the car hostage, watching on in smug bemusement as the woman attempted to shoo him off by shouting, clapping, and flailing wildly. After a few minutes, she dejectedly tossed her purse into a nearby clearing for him to claim. He happily abandoned the car to fetch his trophy, which he took into the nearby undergrowth, never to be seen again. (The bag wouldn't be seen again, that is; I'm sure *he* reemerged when the next tourist arrived.

Another woman stubbornly insisted on bringing her designer handbag into the national park area at the Cape of Good Hope (the furthest tip of the peninsula), despite the repeated warnings of the staff. As soon as she passed through the entranceway, one of the animals, who had undoubtedly been eyeing her since she got out of her car, swung down from his tree and yanked the bag out of her hands, then scampered back into the branches before hopping gracefully onto the roof of the building. He then sat cockily above the group of visitors, with their poor tour guide watching helplessly as the woman screamed at him to climb up and retrieve the bag. Despite being only about a third the size of humans, baboons are powerful, canny, and apparently have a penchant for Gucci accessories – the woman never stood a chance. Her screeching protestations got louder when the monkey got bored of dangling the bag by its strap and decided to show off his fine-motor skills, as well as his impressive grasp of human engineering. He started by unzipping the main pocket of the bag, then rummaged around. On finding a travel pack of tissues, he delicately pulled several out one by one, pausing each time to let them flutter down onto the group. He then found a sports drink bottle, and confidently pulled up the nozzle and tilted his head back while squirting some of the juice into his mouth. Having made his point, he flipped the bag upside down, dumping its remaining contents onto the group. He then scampered away across the roof before leaping off into the brush.

These were the only direct confrontations with baboons I saw, but their presence always loomed. No handbags were safe, this much was clear. According to the park workers, more violent attacks were rare. While this sounded like good news at the time, one of our group outings was such a PR disaster that I was left hoping the monkeys would come put me out of my misery.

SAUSAGE FEST

Looking back, the expedition hadn't started too badly. About ten of us had organized a camping trip to Kommetjie (pronounced *Com-eck-ee*), a secluded national park on the west coast of the peninsula. We arrived in the afternoon and assembled our tents around the firepit in the centre of our rented campsite. I helped set up and rolled out a few sleeping bags. My contributions didn't particularly shoot me up the social rankings, but I was happy to ride with the peloton for now.

Once our tents were prepared, we sat around the fire. Somebody suggested we take turns describing the most beautiful thing we'd seen in South Africa so far. We got halfway around, with people giving the names of their favourite mountains or beaches, or perhaps showing a picture of a wildflower they'd found on a hike. Nothing earth-shattering, but wholesome enough. It was then the turn of Nicolas, a Spaniard who had been openly flirting with a Mexican girl named Isabella since his arrival earlier in the week. He'd been late to the fire and shifted people out of the way so he could squeeze in next to her.

'To me, the most beautiful thing in Africa is ... Isabella.' He gazed at his subject longingly and reached out to brush a loose strand of her hair behind her ear.

'Aww! Oh my God, he's so romantic,' squealed one of the American girls to my left.

'I know. It's like a movie,' said the one next to her.

This spurred our Iberian Romeo on; his newfound confidence prompted him to test the waters by resting his hand on Isabella's lower back. He started looking around the circle with an unctuous smirk, trying to catch everyone's eye. When he noticed a few people on the far side of the fire had been deep in their own conversation, he called over to get their attention.

'Hey guys, did you hear what I said? The most beautiful thing in Africa is Isabella!'

The American girls – the very same ones who had heard him perfectly well the first time – fell for it again.

'Awwwww, he really *means* it!' Incomprehensibly, their praise was *louder* than before. I could hold my tongue no longer.

'Oh, come *on*, there's no way you're buying this. Just look at him! He knows exactly what he's doing.'

This was taken poorly, particularly by Isabella, the jewel of Africa, who glared at me for the remainder of the activity. Well, she *would* have done, I'm sure, but nobody else wanted to take another turn, as Nicolas had 'set the bar too high'. Plus, the sun was about to set, so we decided to go down to the beach and watched as it sank into the Atlantic.

Afterward, as we made our way back along the trail in the dark, I found myself walking behind Isabella. Her suitor, his confidence now unbridled, would periodically run up behind her and spank her playfully before scurrying away back down the trail and out of her sight. She seemed to enjoy their flirty little exhibition, playing dumb and never turning around to catch him in the act. The next time he snuck by me to make another pass, I joked to him that I'd been taking turns with him and spanking her too, but she just hadn't realized yet. She'd seemingly been in conversation, but now instantly wheeled around to scream at me.

'I fucking knew it was you, you creep! You're a piece of shit!'

Thankfully, Nicolas knew I was kidding and restrained her as she started toward me, pointing out that I'd been walking a couple of metres behind and talking with somebody else (who corroborated this). I let him do the talking and managed to refrain from asking why she hadn't turned around even *once* if she'd been so sure it was me. I knew I had no choice but to accept him as legal counsel, but didn't particularly appreciate his chosen line of argument. In skipping straight to a circumstantial alibi, we were conceding that she was right about my character. We certainly weren't ruling out that I *was* 'a piece of shit' who would definitely grope her given the chance, your honour. All we were ruling out was that, in *this* instance, the piece of shit had calculated that it would have been too conspicuous to risk it.

Before things got worse, somebody got a text message and excitedly announced that another small group would soon be joining us. One of the imminent arrivals, a pretty, popular Australian girl, had been caught by a rogue wave while surfing earlier in the day and, in the tumult, had been hit in the head by her surfboard. A few friends had stayed with her until she'd seen a doctor in the afternoon, and they now sent warning ahead to the campsite.

'She was slashed across the face by one of the surfboard fins. Whatever you do, *don't* point it out when you see her.'

I panicked. I'd never seen a victim of a catastrophic head injury before, and didn't know the recommended protocol. *Were you allowed to look at the wound? What if you had no choice?* I steeled myself, reasoning that, if Jackie Kennedy could stay calm enough to fetch a chunk of her husband's skull off the rear of their convertible, I could surely handle whatever was in store.

When the latecomers arrived, we rushed over to their car with bated breath. One by one, the girl's entourage emerged from the car. Finally, the victim herself. I couldn't help it. I looked – I needed to know what we were dealing with. But all I could see was a scratch on her cheek about a centimetre long covered by a tiny strip of tape. *Was that it?* I hugged her and, in my relief that the situation had clearly been exaggerated, forgot the instruction we'd been given a few minutes before. As we separated, I grinned, and gestured toward her forehead:

'Surprised you could make it – looks like you were nearly decapitated!'

I chuckled, figuring I'd handled it well and broken the ice for the rest of the group. But as I heard the gasps from the American girls and saw the surfer reach up instinctively to cover the scratch, I knew I'd made yet another unwelcome contribution. Tyler, a skinny, beanie-wearing dude from Vermont, put his arm around one of the girls and began to console her as if she'd just been rescued from a burning building. Incidentally, it was the same girl he'd been relentlessly hitting on since we'd arrived – *what were the chances?* When he clocked that he was leaving invaluable chivalry points on the table, he seized his chance.

'Come on, bro. Seriously? No *real* man would *ever* make fun of a woman.' After dropping this hammer blow, he glanced down to check if the young, traumatized woman at his side was impressed. It seemed like she was, so he tacked on a pious conclusion.

'Jack, my friend, there are some things you just can't joke about. Maybe you'll learn that one day.'

I didn't respond. I was outnumbered, and we were deep in the wilderness. Besides, I couldn't be too critical: having seen how well Nicolas's theatrics had been received, my 'friend' had adapted on the fly, something I'd failed to do. It was only fair that I take my lumps.

Luckily, Nicolas had brought a guitar to augment his courting efforts and was eager to corral the group back to the fire where they could provide an audience. He wasn't particularly talented and knew only one chord, a shortcoming he disguised by strumming that one chord at varying intensities and tempos. I think he tried to play Wonderwall (the crowd pleaser by Oasis). Artistic flaws notwithstanding, he did manage to distract the crowd. I snuck away during the second song (Wonderwall, again) and walked over to the bathroom building. It was a good thing I'd already heard him play, and knew the source of the garbled, tone-deaf screeches was human. I felt for the park's other campers who could take no such reassurance – it must have been bone-chilling. (It sounded, if I had to approximate, as if one of the nearby baboons was getting disembowelled alive.)

Speaking of intestines, it was time to cook some. Given South Africa's rugged, homesteading spirit and tendency toward a meat-heavy diet, simple, outdoor barbecues known as braais, are an integral part of the culture. As if to emphasize a disdain for culinary pretension, the most popular fare is the spiral-shaped *boerewors*, which translates as 'farmer sausage'. We'd fished one of these from the bargain bin at a small grocery store on our way to the campsite.

It should have been obvious that the sausage's days of *not* being a haven for bacteria had passed. Had they known what was in store for them as they waited in that seedy mini-mart, the little eukaryotes would surely have been rubbing their flagella together with glee. That said, I'm not sure how much the meat's expiration date (a week prior) mattered; we'd never cooked one of these before and definitely botched the job.

Lack of preparation was the main issue; only by the time we sought to make dinner did we realize that our kindling had been exhausted. We peeled the sweating tube from its sheath of oily cling film, then laid the spiral on the grill, watching the pathetically chaste flames lick the underside of the coiled wiener. Before long, there arose another cause for concern – the meat began to swell. We emergency-airlifted the patient off the grill and made several small punctures along the tube's outer membrane to release the pressure. After laying it back down over the flames, we saw that the bloat seemed to have subsided. But within minutes it became clear that this was nothing more than a dead-cat bounce: an olive-coloured puss had started oozing from the holes. But we couldn't afford another airlift; the first had already cost us precious minutes. Someone broke the worried silence.

'Wait a second, don't they mostly *smoke* meat here, anyway?'

'Good point,' came a response. By a majority vote, we elected to persevere, hoping the lack of heat would be made up for by prolonged exposure to the tendrils of smoke that rose from the embers. We let it sit for an hour before coming back. Of course, by leaving the sausage uncovered and keeping it just warm enough, we'd created a perfect breeding ground for the bacteria. This thing was closer to a bio-weapon than a hot dog; I'm not sure even the local baboons would have gone near it. Yet we saw a couple of charred grill marks and took these as sufficient proof that it was edible. I don't usually eat red meat, which was inconvenient in a country with so many exotic options. Having already passed up ostrich burgers, springbok carpaccio, and warthog steak, I felt left behind and didn't want to miss out on my first authentic braai experience. I poked and dissected the sausage until I found a morsel that appeared to be cooked through and nibbled off about a teaspoon's worth before giving up. It tasted fine, I guess, if a little gamey.

As we settled in for the night, a hiker, a brawny Liverpudlian fellow, came past our campsite on his way to his tent and stopped to chat. He offered me a small packet of shortbread cookies (which I scarfed down as a palate cleanser), before casually mentioning that we'd better make sure to keep our fire alive through the night, as well as put any of our food waste in the special, extra-secure bins at the main building. The baboons were watching us from the surrounding woods, he explained, and had been known to rip up campsites in search of food. Before striding off into the dark, he reminded us that they ate meat and estimated that their canines were about two inches (five centimetres) long.

Given the circumstances, these little fun facts horrified us. We weren't familiar enough with the animals to know they don't attack unless threatened, so it was easier to assume the worst: that we wouldn't last the night. As our vision worsened, the thickening darkness left a vacuum for our imaginations to fill. With the offshore wind unremitting, it was impossible to tell what was causing the rustling of the surrounding vegetation. As we sat around the fading campfire, I scanned every bush and shrub, imagining a militia of well-drilled baboons spread tactically and hiding behind cover. The temperature was dropping quickly, however, and we soon had to retreat to our tents.

I awoke a few hours later, wracked by agonizing stomach cramps. There's a chance these were psychosomatic,

I guess, a delayed reaction to my public disgrace, but I think it's more likely they were caused by the expired meat. My digestive/immune systems had resisted the toxins for as long as they could, but they were now in an all-out war, and the crippling pain prevented me from falling back asleep. I sat up and looked around. Nobody else seemed to be suffering any adverse effects. I then realized I could no longer see the fire through the thin polyester of the tent wall. I unzipped the entrance and scurried over to blow on the embers in the hope of resuscitating the little flame which, for all we knew, was the only thing standing in the way of our gruesome deaths.

I sat for a minute before succumbing to the pain, rolling over into a fetal position in the dirt. It was freezing cold and the wind had picked up, but I'd left my blanket in the tent and could barely move. I could hear the trees groan around me and the waves crashing on the rocky shore nearby, but could see almost nothing beyond our small ring of tents. *What a miserable way to go.*

In my feverish paralysis, I wondered if this might actually be an opportunity to salvage my reputation. At this point, I knew run-of-the-mill forgiveness was off the table – Tyler and Isabella would see to that. My only option was to overshadow earlier events with a more compelling narrative: martyrdom, for example, *that* should do it. To that end, the savage baboons lurking around the campsite could yet prove valuable. All there was left to do now was wait. I drifted in and out of a fitful half sleep until sunrise. Alas, no torch- and pitchfork-wielding crucifixion party arrived to take me away. My stomach, albeit bruised and tender, seemed to have come through the worst of it. Survival meant I had to travel home with the others. On the way, I heard Isabella whisper to one of the girls that she was sure I was lying about my stomach; I'd actually been creeping around, trying to get near her tent to hear her and Nicolas having sex.

My faux pas pretty much cemented my status as pantomime villain; I was the perfect foil for other guys to exhibit their virtue. I put up with it for as long as I could, but the situation soon became untenable.

BLUE BALLS
(OR: ANY PORT IN A STORM)

Unfortunately for the other guys, the conquests of Nicolas and Chet proved anomalous. Despite their increasingly shameless efforts, the release they so longed for seemed unattainable. This was usually blamed on bad timing: the target/girl in question had (they swore) flown home *just* before their emotional spark could be consummated. Another excuse was the housing infrastructure which was, in fairness, particularly inhospitable to any couples seeking to become 'one flesh'. This was especially true in the larger, co-ed accommodations, where bedrooms, bathrooms, and common rooms were all shared.

Several of the more libidinous couples, however, were unperturbed. In at least two houses, civilians were instructed to stay in their bedrooms after curfew, as the common rooms had been reserved by these couples for conjugal relations until morning. For the young men forced to share a bedroom with five others, having to fall asleep to the sound of the dreadlocked white girl getting railed by the chubby pothead from South Carolina (who exclusively wore Looney Tunes-themed pyjama pants, and went by the name of Bones) would have done little to alleviate the tension.

This venereal logjam meant the boys' bunk rooms became incubators for an aching, feverish desperation. The situation was so insidious that it even subsumed newcomers, who ought to have been carefree. This made the situation more difficult to take seriously, and no less insufferable; especially during social gatherings. Although my interactions with those living in the bigger houses were infrequent, and invariably brief, it was clear I was still seen as a threat nonetheless. Whenever I found myself in casual conversation with a girl, it was never too long before the guy who'd evidently claimed her would barge over and interpose himself, slightly out of breath from fighting through the crowd.

Two Mexican guys were particularly territorial. I tried not to take this too personally; we shouldn't, as the saying goes, attribute to malice what is adequately explained by stupidity – or, in this case, by horniness. Still, I only had so many cheeks to turn.

The final straw was laid as I was visiting their house one Friday evening. A few of us were discussing whether we ought to walk or take a taxi to a local restaurant. If we wanted to go on foot, I told them, we'd be better off going tomorrow afternoon – it would be safer with more daylight. One of the amigos overheard me from across the room and came over to square up with me.

'Hey, man. You know what we call you? *Preocupado* – it means *worried*. You think too much. You got to understand – this is Africa. Yes, there's crime, there's violence, there's dangerous animals. But who cares? You have to relax, or you won't enjoy it!'

I knew there was a grain of truth somewhere in this advice, so I took the nickname as a compliment (plus, it sounded exotic). And I knew none of this was his fault, caught, as he was, in the clutches of something far stronger than himself. But as I watched the smug little bastard gleefully querying the girls as to whether they agreed, before (I shit you not) launching into a story about how his father was a best friend and business partner with Carlos Slim (Mexico's richest man, we were reminded – *twice*), I knew the Rubicon had been crossed. *Things just got personal, compadre.*

I had to respond. But how? For the moment, there was nothing I could do. Then, about halfway through my trip, we were dealt a wild card.

The first sign something was amiss was when the organizers sent us a memo requesting that we 'please, try to be understanding' with a volunteer who was scheduled to join us later that week. *'Ravesh has been somewhat difficult to communicate with. We think he might have some problems with his confidence and social skills.'*

This was the first time they'd ever made such an appeal, so we were unsure whether it was a well-meaning plea or a disguised warning. The answer soon became clear when several of the girls confirmed the assessment – well, technically only *half* of it. Any shyness the man suffered from was, apparently, entirely situational. A couple of weeks earlier, he had joined our online group where we connected with each other and could share pictures of our trip so far. As if it was a mail-order bride site, he'd picked his favourites and reached out to them, detailing how excited he was to see them in person. He didn't know, or didn't care, that these individuals lived together and would inevitably find out that he'd been soliciting eight of them at once. Until now, they'd assumed this character was one of the guys, or maybe an ex-volunteer, playing a prank.

He had listed his birthday as 1 January, 1978, and his location only as India. The latter part, we supposed, could well be true: his messages, clumsy, blunt, and concise, certainly read as if English wasn't his first language, and both of his pictures displayed what looked like an Indian man. But *this* was another problem entirely: while the pictures corroborated his ethnicity and birth year, they were, indisputably, of two different men. The programme managers didn't seem to mind.

'Look, he says he wants to come and take care of the children. We don't want to turn away those who want to help – it might discourage others. We don't get many people from India. Are your concerns anything to do with where he's from?'

It was obvious, of course, that it was really just the man's *money* they didn't want to turn away. They'd done nothing to disabuse him of his hilariously obvious fantasy that he'd be welcomed by a harem of giggling, submissive virgins. This was profit-seeking masquerading (like it so often does) as morality. But, with the race card having been played, the issue was settled—for now.

This left us with the million-rupee question: where would he be housed? The two smaller houses were girls-only, while my little apartment had room for just four guys. The three bigger accommodations were co-ed, this included the main building (the one that had been broken into). It was here that the coordinators chose to place Ravesh. The fact that it was essentially a hostel, they said, should allay our concerns. The large number of guys on hand would act as deterrents and, if needed, *enforcers* were any tomfoolery to occur. To my ear, this all sounded suspiciously similar to the mob-justice protocol we'd learned from Charles, but I was thrilled all the same. Nothing against any of the girls, but I didn't know them too well. The people I *did* know, and had a vested interest in pissing off, were our aforementioned *desperados*. I cannot tell a lie, the thought of laying an Ravesh-shaped grenade at their feet was thrilling.

And this is how I found myself siding with the programme administrators, saying things like, 'Innocent until proven guilty', and 'Come on, let's all just take a breath. We [the volunteers] are supposed to be society's most open-minded, its most generous, aren't we? We've gotta give him a chance.' Looking back, it's telling just how much of a blind spot guys have regarding women's physical safety. The state of vigilance I maintained down in South Africa was tiring. Imagine having to spend your *life* in such a state, no matter *where* you are. Guys may sympathize with this at a superficial level, or even experience it in certain situations, but we can't internalize it in the same way, which means we can be pretty crass about this sort of thing. Unfortunately, at the time I was far too blinded by competitiveness to worry about anything except screwing over my *amigos*. In my defence, I know I wasn't the only one being selfish in the (Proxy) War of Ravesh. The other guys may have happened to side with the girls, but it was obviously just a happy coincidence, a temporary alignment of interests. What *wasn't* a coincidence was that the Venn diagram of 'girls they wanted to protect' versus 'girls they wanted to inseminate' overlapped into a near-perfect circle. (Such is the case with any guy who's a little too happy to talk about how much of an ally he is – it's the oldest trick in the book.)

Anyway, with Ravesh's arrival date less than a week away, the atmosphere was fraught. The guys were more panicked than the girls, which I enjoyed; the plan was already working. But I'd made a fatal mistake: I'd underestimated my opponents' condition. By even suggesting that I sympathized with Ravesh, I was seen as his accomplice, his right-hand man. Previously, I'd only been an annoyance. Now I was an existential threat. In their eyes, I had committed high treason. The charge? Cock-blocking – on at least a dozen counts. And so I started to second guess myself. *Had I been too rash?* I'd hitched my wagon to a rogue agent over whom I had no control, and who didn't even know of my existence. He had no idea what he was walking into. One misstep and we were done for.

The good news was that all hope wasn't lost. The court of public opinion had largely made up its mind, but I knew the jury wasn't unanimous. I'd noticed a few of the girls echoing my calls for patience, even telling off other girls for being too sensitive. Later on, I did some digging and learned that none of them had received solicitous messages from the applicant himself. Included in this cohort was Isabella, of all people – not exactly a Soul Sister of mine. But that didn't matter right now; I was in the coalition-building business. You've got to be able to put your differences aside in service of a shared objective. Before we got to any of that, though, we had to stop the bleeding. Specifically, Ravesh had to quit sending his chronically horny messages. That meant no more talk of anyone's 'big bobs,' or 'sweet vegene'. If he could just lie low for a while, I could start drumming up support …

Nope. Horny is as horny does. Having taken the hint that the first spate of girls weren't interested, he was now making his way down the list, with one or two new girls being contacted every day (their feigned disgust only barely disguising their relief for having passed muster). Gradually, they jumped ship in favour of the anti-Ravesh movement.

To make matters worse, just hours before he was scheduled to fly from India, we noticed he had changed his photograph online – it was now a *third* different man. Not a great couple of days at this public defender's office.

৯৯

I didn't meet our silver-tongued devil for a few days, since he'd been placed in a different school. I was glad to hear that he'd assimilated well with his housemates, but these good tidings only lasted about a week before reports began to trickle in from several of the girls, who claimed that he'd been hanging around in the hallway while they were showering. He'd realized the girls had to pass through the corridor (usually in just a towel or their underwear) on their way back to their room. Sometimes, he'd pace back and forth, pretending to be taking a phone call. Or he would pretend to have been wandering through the hallway on unrelated business.

While the girls seemed not to appreciate these 'spontaneous' encounters, I couldn't help but be impressed by the degree of logistical difficulty and risk he undertook; how meticulous his choreography had to be. If there were even *one* independent eyewitness to his shenanigans, the jig would be up.

After the break-in, we'd heard that management had installed CCTV cameras in the main house. I had no idea how many there were or where they'd been placed, but I was desperate to find their archives – the footage would surely have been riveting:

Perimeter Cam 4: Our suspect lurks outside the single-story house, kneeling in the grass under the bathroom window until he hears the shower turn off, at which point he begins counting down in his head from ninety (he's studied the timing of her routine exhaustively). He stands, brushes himself off, and walks coolly through the house toward his target.

Cut to Hallway Cam 2: He arrives just in time, waiting against the wall just outside the bathroom. He tries to appear nonchalant. The footage, though grainy, seems to show him pretending to yawn. He repositions himself against the doorframe. She stands beneath him, shivering from the fear and the chilly air hitting her wet skin. He knows he only has a few moments to gain her confidence. Luckily, he's rehearsed his pitch:

'Wow, I had no idea you like to shower every day for between five and seven minutes between seven-thirty and eight o'clock in the morning! That's my favourite time to shower, too. I have so much energy after sunrise. I feel like it's my time to make my mark. To seize the day, know what I mean? That feeling of: Watch out world, here I come! I'm going to have my way with you!'

He gestures to the puddle she's now standing in.

'Oh no! You're dripping all over the floor! Is that the bath-water, or are you just happy to see me? Ha. Kidding, again.'

(Of course, as the low-budget CCTV offers only silent footage, the specifics of any verbal exchange are mere speculation.)

Frankly, I was kind of surprised that Ravesh was the first man to attempt such an elaborate manoeuvre, given the exhaustive lengths these guys were willing to go to ingratiate themselves. This included the instant adoption of any opinion, personality trait, or aesthetic for which the girls indicated a preference. If even a handful of the girls had declared a fondness for guys who wore earrings, for example, I have no doubt that every jeweller and piercing studio in town would have been fully booked for the next week. This all sounds cynical, I know, but I'm a romantic at heart. Ravesh's underdog narrative was certainly compelling in its own right, and I hoped he found true love wherever and whenever it presented itself …

In my *perfect* world, however, his wildest fantasies would come true while he was still in town. As far as schadenfreude went, the thought of the Mexicans finding Ravesh elbow-deep in one of 'their' girls was, frankly, irresistible. There would be simply no coming back from getting cuckolded by the barely literate middle-aged Indian man wearing stunner shades and a pit-stained, knock-off Polo. I may be *preocupado* now, but those spoiled meat-heads would be *preocupado* forever.

ôâ

One evening, out of the blue, a few of the guys at Ravesh's house invited me over. I agreed, foolishly assuming they wanted to bury the hatchet. But the opposite was true: this was a summons to a kangaroo court.

When the topic arose, I did my best to defend my Hindi client. I suggested, gently, that perhaps they had performed a calculus, consciously or not, and concluded that their chances of impressing the girls would be higher if they ganged up on the heavy-handed newcomer. In my view, theirs seemed like an equally cynical but far hornier version of Pascal's wager: we might not be *certain* that kicking Ravesh out of the house will help our cause, but we're damn sure it can't *hurt*. Surely,

I argued, we'd rather prove ourselves in a more legitimate way. Besides, would any of us really want a woman so gullible as to believe such a transparent ploy?

Yes, it turned out. A resounding *yes*. The guys were so desperate to improve their standing that any credit they stood to gain, however microscopic, however relative, was gold dust.

I was running out of ammo. It was time to try something a little less abstract. I wheeled out the 'bro code' (i.e. that dudes should look after each other, and not let girl-drama get in the way). I pointed out that Ravesh wasn't the root cause of their frustration. He was just a scapegoat. Our true enemy was the ruling class we so blindly served. It was time to face facts: the girls had no interest in hooking up and were clearly just revelling in all the attention. To keep the flattery and deference rolling in, they understood that they needed to keep as many 'hopes' (read: boners) alive as possible.

The epitome of this was one girl who, after a couple of months, had moved on from Cape Town to work with a different organization based in Namibia that focused on animal rescue. After just two weeks, however, word got out that she planned to make a miraculous return. Explaining that she missed the group dynamic she'd grown to love, she'd cut her time short looking after the orphaned, crippled elephants. Sadly, all the female friends she'd made in Muizenberg had gone home. Coincidentally, the only cohort remaining from her first stint was the group of guys who'd pursued her. But this wasn't some romantic pilgrimage to reunite with a specific lover. Weeks before the drama of her Namibian interlude and subsequent curtain call, one of the other girls had informed me that there was a boyfriend waiting for the girl back home, with whom she'd been keeping in regular contact. All of this, of course, was completely unbeknown to her devotees in Africa.

I suspected that the other girls, having seen the servility on offer, were withholding similar backstories of their own. And who could blame them? Instead of continuing to debase ourselves, I knew our only hope was to hold the line until the girls were required to take even the slightest initiative.

If I'd triggered any enlightenment, there was no evidence. They heard me out, but it was clear they'd decided their course of action long ago. In their estimation, this little 'multiculturalism' experiment had failed, and it was time to make things right. They'd already made Ravesh pack up his things and had escorted him out of the house, before sending him off in the direction of my apartment; my room-mates and I were to let him move in. I ran to the door, planning to chase Ravesh down and send him back to his harem, but two of the bigger guys pulled me back by the shoulders. As they moved in front of me to block my exit, they pre-empted my next question.

'We've already talked to the coordinators. They're leaving this problem for us to deal with. We can't let you leave until you agree to let this be the end of it.'

I wondered where all of the girls in the house had gone, half-hoping I'd be rescued. But this had all been orchestrated, of course, and no reprieve was forthcoming. Mob rule was being brought to bear. I turned to see a few more of the mouth-breathers standing behind me. De-escalation was, I feared, going to be difficult. Weeks of carnal longing had dulled activity in their frontal lobes. Things like nuance and humour had been cut adrift from their collective consciousness, poly-syllabic words were soon to follow. All that remained for these sexually engorged brutes was the primal, incessant desire: to *rut*. In preparation for a sudden call to action, every excess millilitre of blood was being shuttled south to the low-level tumescence they sustained around the clock.

The circle of slightly sweaty and very hormonal young men converged around me, all enmity between them having been forgotten. In these extenuating circumstances, rivals now united to combat their shared enemy – Ravesh, and by exten-sion, me, his Atticus Finch. They hadn't yet made explicit their plan of action, but I could figure out the gist: when you've been turned into a hammer, *everything* looks like a nail. I desperately wished I had some muscle-relaxing Swazi on hand to prepare for the trials ahead.

And yet, however premature my demise felt, I took some solace knowing it was All Part of His Plan. That's right. In sparing me from the baboons during our camping trip, the Big Guy Upstairs had been saving my body for a far more sacred bacchanal. I, like Jesus before me, would assume the role of sacrificial lamb, offering myself as tribute to the trem-bling, priapic horde. *This* was what being an ally really meant:

taking one for the team. It was my duty – nay, my *privilege* – to preserve the chastity of the girls. *Oh, man; they were gonna owe me, big time.*

Still, there remained the question of my stigmata; my demise wasn't going to be as photogenic as Christ's, that much was clear. The way the forecast was looking, my ass would be getting turned inside out, left as a scarlet, glistening knot, not dissimilar to a female baboon's during mating season. My days of continence were well and truly behind me; I'd be leaking a snail trail behind me for weeks. I could only hope the frescoes, statues, and devotional candles would depict my wounds tastefully.

My thoughts were interrupted by my phone ringing. I pulled it out of my pocket. *Chet.*

'Hey, dude,' he said, 'Ravesh just showed up in front of the house. Looks like he's got all his bags. Know anything about this?'

'Oh, uh, yeah! He's just staying with us for a few days. A different volunteer arrived last minute so they had to shuffle the sleeping arrangements around a little.'

'All right, whatever. See you later.'

Chet bought my hostage-at-gunpoint story and had already hung up, but I knew this was a lifeline I couldn't afford to relinquish, so I kept the phone to my ear, giving the group a conspiratorial wink and a thumbs-up, as if to say I'd be back with them shortly.

'What was that, Chet? You can't find your keys, and I'm the only one with the remote to open the outer gate? So I should hurry home right now? Shit, OK. Well, I won't be long, I'm only at the main house. See you in a minute.'

My captors reluctantly let me out and I scurried home, making a mental note to come up with an emergency safe word with my friends in case we ever needed to inconspicuously ask for help, like those bars where women can order an 'angel' shot if a date's going badly.

Thankfully, Ravesh's stay with us was brief and uneventful; living only with other men rendered him mostly toothless. Apparently, though, volunteering alone didn't scratch the itch. In the middle of the following week, he disappeared abruptly. The group coordinators insisted that he hadn't been ejected from the programme and had left under his own volition. When we checked his Facebook page a few weeks later, he'd uploaded a picture of him with what looked like a wife and three young children. We did not reach out for clarification.

BROTHERLY LOVE

Chet left for home soon after, ready to rekindle things with his muse back in New York. Right before he left, he inexplicably gifted our bike to the young daughter of the family who lived downstairs. I only learned of this when another volunteer showed me a picture that the girl had posted online, posing with her 'New Bike!' Since Chet was already on his way to the airport, it fell to me to repossess it. I called him that evening before he boarded his flight, but he played dumb.

'Sorry, bro, didn't think you'd care.'

This was a lie; the bike provided vital transportation in an area that was dangerous for lone pedestrians, and he knew I still had a month left in the country. I managed to reclaim it, but he got the last laugh. The next day, he posted (without crediting me) the video I'd captured during the shark dive; within a few weeks it had been viewed by thousands of people.

The guys who arrived to fill the apartment were much more agreeable. One of the guys, Jamie, from England, had just arrived from South-East Asia. He'd enlisted in the navy the year before, but his deployment had been pushed back, which had given him a few months to kill, during which he'd decided to travel the world. By the sounds of it, he'd been getting his money's worth: 'Thailand was amazing, I was like a god there. It's great – whatever happens, you just go to the pharmacy the morning after. I've got to take a pill for the next few weeks that makes my pee look radioactive, but other than that, I'm totally fine. Small price to pay.' (I'm positive he would have shown me had I asked; I now regret not doing so.)

The others were Fritz, a sporty, cheerful German, and Peter, a bearded, slow-talking Hawaiian. Our little clique seemed harmonious enough. As the elder statesman, I showed them around – we visited food markets, went to the beach, and did a couple of hikes. They were keen to see the city itself, where until now I'd spent no more than a few hours. Other than my walking tour early on, I'd really only seen it from atop the surrounding mountains.

❧

The city proper has a few distinct districts, including the tourist-centric Waterfront, the central business district, and the historic section, with Greenmarket Square and the nearby Slave Lodge (now a museum). The rest of the City Bowl comprises quieter, residential neighbourhoods with difficult-to-pronounce Dutch names like Oranjezicht and Tamboerskloof.

With so much of the city having suffered crippling demolition or economic abandonment – the aforementioned District Six, for example – the only non-white neighbourhood that's of any interest to tourists and has kept its own distinct identity is Bo-Kaap, which is inhabited mostly by Cape Malays, a majority-Islamic faction of the coloured population. The modest, two-storied homes within these few city blocks are famous for their striking façades of bright or pastel yellows, pinks, and blues. During apartheid, this was designated a racially segregated area. Now, this sleepy little neighbourhood occupies prime real estate, although it's been designated a protected heritage site in the hopes of staving off gentrification.

Although the houses here had their distinctive colours long before the advent of social media, this practice was unwittingly tailor-made for a world in which the sole intent of many travellers is to generate online content. As such, whatever legal protection the neighbourhood has with regard to its zoning does little to mitigate the unremitting flow of foreigner-laden luxury tour buses (the same ones that hurtle back and forth down the length of the peninsula to the Cape of Good Hope). Mercifully, the narrow streets are almost impossible for these buses to navigate, so most of them park a few blocks away, leaving their hordes of selfie-stick-brandishers to spill into the streets.

The residents watch on wearily as they're filmed and photographed. Not intentionally photographed, mind you. Us tourists don't bother talking to any of the locals, you see, let alone ask for their permission. If they didn't want our attention, they should've thought about that before they dressed their houses up so nicely. Thankfully, photography apps are now so advanced that they allow for comprehensive editing on the go. As my room-mates and I walked around, I saw one girl scrolling through the glamour shots her boyfriend had taken. She found one she liked, but lamented that the background had been 'ruined'. In his carelessness, her boyfriend hadn't noticed an old man looking out his front window. Luckily for the boyfriend, the girl was able to edit the man out in seconds, restoring the picturesque fuchsia backdrop to its full, untainted glory. (She then turned the airbrush gun on herself.)

÷

The guys and I usually relaxed in the common room after work. One Friday afternoon, we'd bought a case of beer. After finishing his first round, the Hawaiian made his way over to the refrigerator. I'd finished mine, too.

'Hey Pete, could you grab me a beer while you're up?'

'Sure thing, dude.'

His skater-bro drawl was exaggerated, and his parlance included turns of phrase like, 'That's rad, home-dawg,' and 'Could you flip on the 'Fi, brah?' ('Fi as in *Wi-Fi* – during their first few weeks, I'd invited them to share my internet hotspot.) He never broke character – it was as if he had some sort of high-functioning aphasia.

'Hey, *Pehter*, could I have one too?' Fritz asked.

'Yeah, but that's not my name. It's *Pee* – ter.'

'Oh, sorry. It's difficult for me to say it that way. And I have a friend from my home town with the same name, so if I pronounce it *the German way*, it is just habit, not intentional.'

'Look, bro. I don't care. That's not my name.'

Fritz was visibly embarrassed. Though he had a noticeable accent, his grasp of English was impressive, and he'd always been lighthearted and self-effacing about any misunderstandings or mistranslations; there were certainly no grounds for us to be critical. I looked across to the neon-urined Englishman, whose bewildered expression mirrored my own. We waited for a few seconds, but no 'Ha – gotcha! I was just fucking with you,' moment arrived. I tried to intervene.

'*Pee* – ter, *Peh* – ter. Potayto, potahto. Who cares?'

Jamie followed suit.

'Yeah, come on! Bring me a beer, *Pehter!*'

I caught the German's eye to show him we were on his side. Laughing with relief, he joined in.

'Three beers please, Herr Pehter!'

We watched the Hawaiian in nervous anticipation. He returned from the fridge and set the bottles down on the table before sitting down and leaning back in his chair. He appeared to relax, closing his eyes and sighing, before resting his nose on tented fingers which he held before his face. For a moment, I was optimistic. After all, just minutes before, this shaggy, puka-necklaced guru – at the ripe old age of thirty, ten years our elder – had been espousing the virtues of Buddhism: the inner tranquillity that comes from a life spent in pursuit of mindfulness and simplicity, the abandonment of earthly desires, etc. But the uneasy silence lingered. The American took off his Rasta beanie and set it down calmly, then pushed his chair back and stood up. He calmly strode around the table and stood over Fritz. Without saying a word, he leaned down, drew back his fist and struck the boy in the sternum with a vicious jab. He then straightened up and returned to his seat.

'What the fuck!' I said, rushing over to check on the wheezing victim, who'd crumpled to the floor, completely winded. 'He didn't mean anything by it. We were just messing around. You can't just do shit like that!'

But, like his beanie, Peter's nonchalant façade was already back in place:

'Don't worry, bro, let's all just chillax. It's all good in the 'hood.'

Bizarrely, it *did* seem like he was telling the truth, at least in his view: he'd been presented with an issue and had acted swiftly and decisively to resolve it, disciplining the boy with mechanical, deliberate precision. Apparently, there are some lines you *just don't* cross, and someone had to be held accountable (bro).

Pete's chillaxed vibes notwithstanding, it took the rest of us a few days to recover emotionally. I'd witnessed few interactions so bizarre. These two men had been friends, or at least friend-*ly*, from the moment they'd arrived. Was this how things worked in Hawaii, perhaps? I'd never even gone as far west as California. The only one of us who'd explored that side of the world was Jamie, whose souvenir was his unfortunate scarlet – well, *neon* – letter. (And who knew what else he might be carrying.)

Maybe these were the sorts of growing pains we had to come to terms with in the modern world. There's a lot of different folks all sharing the same space, and it would be naïve to think we'd all immediately be compatible – it was going to take some work, some compromise. That was what I was supposedly here for, after all. I couldn't have it both ways, complaining about it feeling like a sorority house, yet also closing myself off to the unfamiliar.

We may have been too in shock to appreciate it right then, but there *was* something miraculous about this confluence of events. What a melting pot this trip had been. We had Jamie, the Englishman, taking shady eastern medicine for whatever venereal disease he'd contracted from his tryst(s) with the Thai or Burmese ladyboy(s). I'd also met Aaron, the French-speaking Ghanaian from Canada who played American football. And Chet, the well-endowed, English-nosed, Jewish New Yorker who'd come all the way here with his Austrian-made kite-surfing gear. I'd watched a Spaniard woo a Mexican girl with the music of a band from Manchester. I'd smoked weed (or at least something resembling weed) grown in Swaziland, I'd tasted mushroom extract from god-knows-where, and I'd knocked over a hookah pipe shipped from somewhere in North Africa, spilling its coals onto an Australian girl who was being courted by two Mexicans. On my inbound journey, I'd sat next to a middle-aged man who'd never been on a plane, let alone to another country, until just weeks before. And, lest we forget, there was Ravesh, an Indian man who, thanks to the magic of the internet, managed to leave his small village and explore a continent that he would likely never have been able to visit otherwise. To cap it all off, we now had Peter, a blue-eyed Pacific Islander, a practising Buddhist with a penchant for Jamaican reggae, punching a German as we all drank Danish beer.

This was what was possible now. Infinite combinations of peoples, cultures, and ideas. So we've got to be able to take the bad with the good, and to apply some sort of cultural relativism. Perhaps I, with my elitist mid-Atlantic sensibilities, shouldn't be so quick to judge the behaviours of an islander whose little archipelago had only been dragged into the US a few decades ago. For all I knew, this could have been a vestige of some tribal rite of passage. I'd heard of the Sambians, a tribe in Papua New Guinea, where the elder males argue that women are inherently evil and have the power to strike men down with terrible diseases. Thus it's crucial to inoculate the vulnerable prepubescent boys from all this pernicious influence. Thankfully, the older men have an unlimited supply of the necessary vaccine. (No points for guessing – it's semen.) This sort of thing may sound crazy to you and I, but we've got to remember that other cultures have all sorts of different ideas about adulthood and social hierarchies and so on.

The only other explanation for Peter's apparent overreaction was that there existed a fifth noble truth of Buddhism. A lodestar so illuminating, so pure, so self-evident as to be ineffable, with any attempt to record it considered blasphemous. This truth, I surmised, was roughly: *Thou shalt not struggle to pronounce a word in thy second language.* Any infractions were to be punished swiftly and with extreme prejudice.

It was also entirely possible that our guru was simply full of shit.

ઢ

Fortunately, we were soon presented with a couple of opportunities for group bonding. Jamie had seen a brochure advertising a paintball battlefield nearby and signed us up. We were all placed on the same team, which was unfortunate for the German, as I'm sure he would have taken great pleasure in sinking a few rounds into his Hawaiian assailant. (A hacky scriptwriter's dream: 'Say *Guten Tag* to my little friend!', etc. Or a reference to Pearl Harbor, even.)

An even better opportunity arrived with the next week's groceries. The administrators inexplicably provided us with several massive pallets of eggs – at least twelve by twelve. Our small refrigerator had no extra space, so we kept them on the floor by the entrance. After a week, we'd only eaten two dozen or so of the eggs between us. As Jamie fetched some to make his breakfast one morning, he noticed a small sticker on the side of the package. No wonder our coordinators had been so generous: the eggs had expired two weeks ago. At this point, the only fitting use for them was recreation.

There was a four-lane motorway about fifty metres away that divided our neighbourhood from several acres of undeveloped land and, a kilometre or so farther inland, a township. From our porch on the second floor, we had a perfectly clear sight line to the road. Given a magnificent heave and the proper arc, the eggs would explode, appearing to almost vaporize on impact with the tarmac. When we were sure there were no cars or pedestrians around, we had competitions for distance, accuracy, and loudest splat. We had a few scares: given the hang time of the projectiles, it was possible (albeit unlikely) that a speeding car could come into range of the landing zone, especially given that a few service roads were in our blind spot from the porch. Hitting one of the gang-owned taxi vans would have been a disaster, but we managed to avoid them.

We were all competitive and took our games very seriously. But for the Hawaiian, we were pretty evenly matched. Try as he might, he just didn't have the knack for it, but we always made sure to let him win a round or two. Worth it, I think, to avoid getting punched in the solar plexus.

BITTERSWEET

My time in South Africa was almost up. I knew how special the last three months had been, how lucky I was to have had so many unique and memorable experiences. I should have been bonding with the other volunteers, making the most of what time I had left. But those last couple of weeks were bittersweet, and in fact some of the hardest. I'd never been this far from home, nor travelled alone for longer than a week. This all caught up to me, and I withdrew into myself. Homesickness is a horrible wall to hit, especially when there's a departure date set and you're desperate for your trip to end on a high. Yet, I found it impossible to escape the malaise. I felt numb, deflated, a shell of myself. I stopped bothering to learn any of the new arrivals' names; it wasn't worth trying to keep up with the ever-changing roster. Although my grouchiness may have seemed performative, it wasn't; my social battery had been utterly spent. This created a vicious cycle, whereby I was painfully aware that I was squandering what time remained, while resenting the others for seemingly having no hang-ups of their own.

Nobody else was ever assigned to my school, which was probably for the best – for the other volunteers' sake as much as my own – I would have been horrible to work with. Besides, once I'd regained my voice after those first weeks, I had the situation mostly under control and, for my pride's sake, I felt obligated to see out the rest of the job. How would I be able to look myself in the eye if I were to put in the hard yards, only to see some schmuck waltz in at the last minute and steal the limelight? Worse, what if it became a cuckoo-bird scenario, whereby my assistant somehow became the favourite? *Horrifying* – not on my watch. *I* was the protagonist here, and I'd already pushed the boulder this far up the mountain without any help. *'Listen, pal, these kids' lives can only be saved by me. These are my roots we're trying to reconnect with; my Mama Africa.'*

Not that I divulged any of this. At every opportunity, I relished recounting my sob story as the overworked and underappreciated public servant. I did derive immense enjoyment from this, but the truth was that I really had become quite protective of my students and was feeling very sentimental. I'd adored my time with them, and I'd like to think this was mostly reciprocated; I'd get dozens of high fives and hugs when I passed them in the hallways between classes, and most of the kids referred to me as Uncle Jack, a colloquial term of endearment.

I spent most of my remaining free time watching tennis. During my last week, the Davis Cup final was on, with Switzerland versus the hosts, France. For whatever reason, my usual spot (the phrenology club) wasn't open, so I walked over to the beach café frequented by the volunteers who lived on that side of the estuary. I watched on as Roger Federer (the greatest player to have ever held a racket) beat Richard Gasquet, clinching the title for the Swiss. During the final set, the table next to mine had been filled by a middle-aged man sitting across from a boy in his early teens – a father and son. They were sitting behind bowls of ice cream. I hung around after the match finished to earwig. A few uncomfortable minutes went by, during which the dad made a few futile attempts at small talk, and their ice creams went untouched. Eventually, he cleared his throat.

'You know, John, this doesn't have to be the end of the world. Everything will be OK. I'll see you on the weekends, and I'm still going to come to all of your cricket matches. Anyway, tell me how you've been getting on at school.'

There was a pleading, tender rawness in his tone; whatever pain the family was going through was fresh. He had clearly been looking forward to seeing his son and was desperately searching the child's face for any sign that he might be amenable to a regular conversation, or even just to feigning one. Not wanting to disturb them or draw attention to myself, I pretended to watch the muted television in the corner, which had been switched over to a close-captioned Nigerian soap opera. I furrowed my brow and nodded along. I knew I was intruding, but I was riveted as I awaited the outcome. But it was nearly time for the dinner rush, a waitress had started taking a wet cleaning rag to the surrounding tables and had noticed me. The bill for my lime cordial (1.5 rand) had been paid and swept away almost an hour ago.

'Sorry, sir, are you finished here?'

Shit. I was being kicked out – unless I wanted to try and argue that I really should be allowed to stay and finish this soap opera episode. I took my time collecting my things before eventually leaving. I peered back inside from out on the pavement, desperate to see how things turned out. Maybe the waitress had spurred them on to tuck into their now-thawed ice cream. But it was getting dark, so I had to head home. There would be no resolution. Not right now, at least.

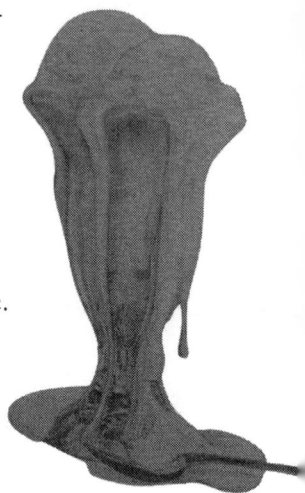

૪ᐧ

The next day at school was my last. Pastor Samuels wasn't there, but this was nothing new. In the entirety of my trip, I only saw him a handful of times. There were days when I would go through my daily routine before discovering him in his office around lunchtime, reading the newspaper or watching a movie on his computer: 'Oh! Hi, Jack. Hope everything's going well out there. Remember, don't be afraid to report those students if you need to. They can be lazy!'

I stopped in to say goodbye to the principal (he gave me a striped varsity tie, which I cherish) before visiting each of the younger classes. I'd travelled over with several packs of crayons and a few dozen little colouring books that I planned to drop off as leaving gifts. At my first stop, a first-year class, I watched one of the quieter kids gleefully unpack the fresh crayons before flipping to a page at random that had a farm scene. He coloured black spots on a cow and used bright pink to fill in the lines on a family of pigs. Incidentally, Elmo was visiting this farm – I'd bought this *Sesame Street*-themed book thinking the brand was ubiquitous, but evidently not as the little boy, assuming Elmo was just any old man, coloured him brown. Once he'd finished, he beamed with pride as he showed me his work. As I looked down into that sweet little face, I felt a swell of emotion. *Was that a lump in my throat, or just my acid reflux acting up?* What a powerful metaphor this was for the value of connecting with those whose worlds differ fundamentally from our own. How uplifting it was to see our cultural hegemony hadn't yet claimed him. So refreshing, so pure … Yet I could feel my indoctrination kicking in. It was my duty to correct him, was it not? *'Listen here, kid, and listen good: Elmo is red, and Elmo will always be red. If you ever pull a stunt like this again, so help me God.'*

In the end, I decided to let him off the hook.

'Great job, little fella,' I said, before moving on to another student.

It was an exam day for the older classes, and after they finished their tests in the afternoon, they were let out for recess along with the rest of the students. I'd brought my camera with me, filming as I walked through the school's hallways. I got lots of group hugs, and I'm even convinced I saw one third-year wiping a tear, although unfortunately this wasn't captured on camera. I wandered around with a couple of the younger kids who didn't want to leave my side. We went over to the other side of the main school building, where the older classes hung out in several small outbuildings. Absorbed as they were in the dramas of adolescent life, these kids didn't seem as devastated by my imminent departure. I wasn't too familiar with this part of the property; I'd never had reason to explore it, as there wasn't space here for any of our sports or activities. The outer walls of the buildings faced away from the school and thus provided privacy from teachers.

As we did our lap, the morning took a less-wholesome turn. I, along with the seven-year-olds, naïvely swung around the corner of one shed to discover a couple sharing a particularly intimate moment. There was a no-fail policy at this school, which meant academically challenged students could be held back as many times as necessary. Some of them had clearly given up: the fellow sitting against the wall had facial hair fuller than anything I was capable of, while the young woman who was straddling him looked older than most of the girls Chet had taken to pound town (so to speak) during his visit. I hurried the little ones away, horrified. I'd exposed them to a traumatic scene, and I wasn't even sticking around to help them deal with their PTSD. Oh, well. That was for the next guy to deal with. And he'd need to find his own whistle, by the way. I was taking mine with me. My work here was done, my mission(s) accomplished.

World? *Changed.*

Self? *Actualized.*

African roots? Consider them *intertwined-with*.

Even so, I opted against getting a TIA tattoo of my own. Going to Africa had changed the course of my life, and that would be enough of a reminder for me. (And it would have been impossible to top the girls' attempt, anyway.) This trip marked the true beginning of my adulthood, representing solitude, discovery, and escape, and proving that departure from the standard trajectory was entirely possible. It had been challenging, of course, but I'd done it. *I'd fucking done it.* But I should have learned my lesson from that bungy jump: *never* celebrate until you're out of the woods ...

The morning I was scheduled to fly home, I went for one last run around the neighbourhood and the marina. I came back and lay down for a nap, only to be awoken by two policemen standing over me.

'We've gotten a few reports from drivers in the area that their cars have been targeted with rocks. Is this anything you know anything about?' My life flashed before my bleary eyes as I imagined being hurled into a cell on Robben Island. After all my brushes with death – and other various forms of subjugation and debasement – I couldn't believe I was destined for such an anticlimactic fate: '*It was the seventy-four counts of egg-tossing what did for 'im.*' This was like Al Capone getting sent away for tax evasion. Thankfully, the officers believed my claims of ignorance and left to continue their manhunt.

I was still on edge as I checked in and passed through security that evening, half expecting Interpol to have thrown me onto the no-fly list before I made it to the airport. When I finally got to my gate, I found an empty seat and collapsed blissfully: I was finally headed back to the familiar world.

As we took off, I peered out of my window to the setting sun. The glowing core was already partway down, but still looked impossibly large, a heat-rippled golden yolk sinking from a sky of bruised purple. It occurred to me that this might be my last South African sunset. I had been so burned out, so focused on the thought of leaving, that I hadn't considered the actual departure itself and what that might mean. The sun started to blur into the clouds, and I realized my eyes were welling up. Perhaps this wasn't going to be as straightforward as I'd expected.

DEBRIEF
(OR: A SUPPOSEDLY NOBLE THING)

To my surprise, by the time I'd been back just a few weeks, all the boredom, loneliness, and angst I'd felt while abroad had ebbed away. This allowed me to properly contextualize, and even appreciate the trip. Hurdles that had felt insurmountable (e.g., South Africa's ketchup and mayonnaise had tasted funny, and their milk was this weird long-life, ultra-pasteurized stuff) now seemed embarrassingly trivial, or were forgotten altogether. In their absence, I was left not with empty space but, curiously, with a sort of filmic negative. Along with the country's obvious upsides – the weather, the landscape, the favourable exchange rate – I found myself desperately missing more subtle and subjective attributes. Its trimmings, smells, flavours, accents. It wasn't about whether these were better or worse than their American equivalents. Indeed, I'd even begun to think fondly of the more counter-intuitive stuff: its idiosyncrasies, its inefficiencies, the 'TIA' stuff. Most important, beyond their being novel, was that these variables had been *present*, providing the staging for so many memorable, formative experiences.

In me, the country had gained a lifelong fan. I now kept an eye out for it in the news, cheered for its sports teams, and recommended a visit to everyone I spoke to. It had suffered, yes, but it simply had too much promise to bet against. Perhaps my soft spot was because I, too, was in a transitional phase: potential was there to be realized, it was just a matter of getting a few affairs in order first.

I thought back to that commemorative chunk of the Berlin Wall. The reunification of Germany in 1990 had, as far as it was portrayed back home, marked a watershed moment – not only for Western Europe, but for the world. A repressive system had been toppled, literally, its once-imposing barriers hacked to pieces and given out as souvenirs. With the spectre of communism defeated, the world could finally drop its guard and open its doors for business: *Come one, come all.* It was only a matter of time before South Africa would dust itself off and sidle up to the global marketplace to enjoy the wares on offer – peace, democracy, and prosperity. I resolved to return to check back in as soon as I felt ready.

It was tougher to assimilate than I'd anticipated. I found myself feeling kind of lonely; my adventure, however special and unique it had been, wasn't really anything that people back home could relate to. I also felt markedly less beholden to the United States than I had before; it was no longer the only place I'd ever lived. This isn't to say that I felt perfectly suited to South Africa, however beautiful it was, and however affordable. This left me in a kind of liminal space. Perhaps this was the danger of spending time in other places: you're always missing somewhere. For now, it seemed like my best bet was to stay light on my feet. I knew this would make finding a tribe (my Soul Sisters, if you will) difficult, but that was fine with me.

୬

This was all well and good, but it was decidedly unhelpful regarding my academic career. I could have changed majors, but this would mean I'd be starting from scratch, making those hideously expensive political science credits I'd earned in my freshman year null and void. *Well, shit. What now?* The system from which I'd broken free had already caught up – and so quickly, too. I'd blown my savings to break away for a few months (the last of which I was too homesick to enjoy), and was already right back home, living in my parents' basement. Being a sub-par volunteer I could accept, but this was far worse: I was also a sub-par nonconformist.

So, although the anti-gap-year soothsayers had been correct, and I didn't have any interest in returning to school, I swallowed my pride and signed up for classes at my local community college. After a semester there, I found a university in Florida that offered to accept my credits (and where tuition cost less than a three-bedroom house). It was in St Augustine, a picturesque little town whose claim to fame is being the oldest in the country, as it was founded by the Spanish long before the pilgrims arrived from England. I loaded my things into my family's spare hatchback and drove the thirteen-hour journey down I-95 in one shot. I shacked up in a tiny apartment right by the beach and enjoyed a balmy, pleasant year, but I never stopped wondering when, and how, I would get back to South Africa.

On one of the first evenings after moving in, I used an online GPS to figure out the exact angle I needed to keep an eye on my pals in Cape Town. (Or, in case push came to shove, and I needed to swim there.) There was an old wooden bench atop the grassy dunes just outside my front door, and for the rest of the year I took my tea or coffee out every morning and evening and gazed out across the Atlantic, my sights set 124° southeast.

෧

The pull got stronger with the Trump's ascent to power in 2015, when America's politics and media descended fully into self-caricature. Admittedly, this was both revealing and, at least for a time, entertaining to watch.

Obama's victories had led to almost a decade of laurel-resting and hearty self-congratulation by the media and most of the country. We'd put Bush behind us and proved that our system worked, that our institutions and values were inherently good. America, as a global brand, was back on top. For Democrats – white ones in particular – the half-Kenyan presented a miraculous opportunity to heal centuries of damage without threatening the comfortable status quo.

Here, the crudeness of our two-party system had finally proved advantageous. There was no uncomfortable grey area, no nuance, only a binary choice. Do you vote *for* racism, or against it? No further action or thought was required.

And then we got Trump. Reports of our democracy's rebirth, it turned out, had been greatly exaggerated. Evidently, our mighty government, with all its checks and balances, its pomp and circumstance, its hallowed halls, etc., could do nothing to stop this spray-tanned septuagenarian moonwalking straight into the Oval Office. *Hang on.* I'd thought we, the world's moral arbiters, the upholders of democracy, weren't susceptible to being taken over by fascists, demagogues, and populists. That was the whole point of our shtick, was it not? Our leaders were the best of the best. So how had this happened?

A lot of the liberal media claimed Trump was simply an aberration, a racist backlash after eight years of a black president. This may have been true to a degree in 2017, but my sense was that they were deflecting blame. Despite claiming that Trump was dangerous and blatantly unfit for office, they'd legitimized him nonetheless, giving him and his campaign free, round-the-clock coverage. When every last one of Trump's Republican rivals – to or about whom he'd said all manner of things – endorsed him, these pundits had scoffed: 'How spineless, how cynical, how undemocratic of them. *Our* side would never debase themselves like that.' They then turned around and wrote puff pieces a few months later after the Donald's first State of the Union address, proclaiming that, to his credit, Mr Trump was looking 'damn-near presidential ... Maybe we shouldn't have been so quick to write him off!'

In light of this, I began to see Trump's victory not as an indictment of him or his party, but of the wider systems that had enabled his rise, and that now seemed completely unable – or unwilling – to stop it. For all the moral panic we'd seen before Trump took office in 2017, the reality was that anyone who was financially comfortable had been able to continue as before; his victory had had no material impact on their lifestyle. The never-ending election cycle was just their version of American football or pro wrestling: pregame predictions, halftime analysis, postgame wrap-up, heated debate over tactics and strategy. An endless source of content to foment outrage and tribalism. It was all just a platform for self-promotion and career advancement, all performed for the mythical 'undecided, middle-class voter'.

As far as his enduring popularity, a factor I *did* believe had some merit was that Trump was a poor person's idea of a rich person. In this regard, he was not an aberration, but quite the opposite: he was the embodiment of the American dream. He was the Übermensch.

The weight of this, I think, can't be overstated, specifically, the extent of our classism. There's racism, of course, both brazen and systemic, but make no mistake: we hate poor people just as much as we hate minorities (even if it's we who are poor). I'd judged the South Africans for their unrequited loyalty to their country's elites, and their willingness to vote against their own interests, but I'd now realized we in the US weren't too different.

The thing is that luck (or so we're taught), has nothing to do with it: anyone with more money or power than us must be proportionally more industrious, wiser, and flat-out more deserving. Our rich don't just avoid scrutiny, but are *lionized*, amassing legions of sycophantic admirers. This is how you get millions of poor and working-class folks, desperate for every proof-of-concept story they can get, pouring forth to defend the character and business acumen of people (e.g. Trump and Elon Musk) who wouldn't even acknowledge them on the street. This also explains how you get those same exploited people arguing that those in power are *justified* in not raising the minimum wage, or providing healthcare, or public transport, or safe drinking water. We believe that if we do not have the 'luxuries' (like healthcare) enjoyed by the rich, we must not deserve them.

The benefit of this is that it allows us to pay little mind when we see other Americans in dire straits. From our point of view, they should be grateful for having an opportunity to build character. We see our indifference not as inhumanity but in fact as a *favour*: 'Let 'im cry himself to sleep, it'll toughen 'im up. He'll thank us later.' The conceit here is baked into our language, so a 'living' – happiness, health, shelter, a family – must be 'earned'. A dignified existence is *not* an inalienable human right, we must toil or sacrifice for it. What these attitudes give you at scale is a society in which the banality of evil is not only leveraged, but enshrined.

ào

Though our buffoonery was, I'm assured, wildly entertaining for outsiders, I'd had my fill. It was time to go; Mama Africa was calling.

But what form would re-entry take? As far as saving the world, my maiden voyage hadn't exactly been a resounding success. Voluntourism, from what I could tell, was mostly just for the benefit of Westerners like me, letting us role-play as Mother Theresa or Bono for a couple weeks before we jetted off back home. Any relief we'd provided those schools, the country, or its children was, I feared, negligible. The school where I'd been placed already had a [nominally] full-time gym-teacher/cleric, and a roster of perfectly qualified and lovely teachers. Sure, I'd shared some beautiful moments with my students, but I'd been bussed into the school's premises from an apartment near the beach every day and had only worked for a few hours before heading home to lounge by our swimming pool. To be clear: my doing the bare minimum is a reflection on no one but myself. By the looks of it, though, most of my peers were putting in roughly the same amount of effort, and the programme organizers seemed to have no problem with this – the occasional sternly worded memo notwithstanding. As long as they could stump up the cash, Ravesh, Chet, or anyone else was more than welcome.

This open-door policy means there's very little regulation as far as editorial control; which invariably leads to do-gooders like me overestimating our contributions. The exotic locales and people we meet or observe on these expeditions are particularly susceptible to this sort romanticization, especially with the advent of social media. Attention-seeking on our own turf is one thing, but using countries like South Africa to act out our saviour fantasies is a little more complicated. The 'let people enjoy things' argument only holds water if the given forms of recreation truly *are* harmless, and our well-meaning but misguided humanitarianism is not.

So, what's going on here? Why is it, exactly, that we feel so much pressure to save the world? Surely, the fact that there are *any* poor or starving people right here in the US and UK exposes every single one of us missionaries (secular or not) as frauds. *Checkmate.*

Or so it would seem. But there's actually a simple explanation for all this. And it's here, I think, that we can glean some useful insight. I'd started to realize that these attitudes, to include our saviour and martyr complexes, weren't a manifestation of opposing, or even particularly different values than those that allowed someone like Trump to become president. It was all from the same petri dish.

For us, you see, there is no contradiction here. *Their* (i.e. anyone suffering outside our borders) problems and *our* problems are not only unrelated, but flat-out incomparable. Their problems, you see, are systemic. Ours are individual, meaning they can – and thus *ought* – to be solved by the individual. *Have you tried getting more exercise? What about keeping a gratitude journal?* The pandemic of loneliness, depression, and anxiety is, we're assured, certainly not symptomatic of any broader failings. *How can our system be at fault if it has given us everything we need to succeed? We all have the same number of hours in a day.*

Besides, a *true* American, be they red or blue, poor or rich, shouldn't even *want* charity, regardless of whether they need it. Americans, lest we forget, are the ones who've been put here by God to pity others. It's manifest destiny. We deal in *pride 'round these parts, my friend.* The only relationship we have with *pity* is as an export. It is we who are the protagonists of the universe, superheroes capable of clearing any hurdle put before us. *Follow your dreams, no matter how unlikely. You can be anything you want.* The caveat of this seemingly uplifting sentiment is that, in practice, it actually becomes more of a directive, a *threat*, the implication being that, since you can be *anything*, you *can't* be nothing. You can't *do* nothing. This leaves us in quite the quagmire: delusions of grandeur, ever-shortening attention spans, and an outsized appetite for showing the world how special we are, but without the tools to channel any of this in a healthy nor productive way.

In light of this, it's really no wonder that we make ideal patsies. Just look at how the responsibility for fixing climate change has been shifted down to us. *You, there! Yes, you! You thought brushing your teeth without turning off the tap was a victimless crime? Wrong. You may as well have held the last*

remaining polar-bear cub underwater and drowned him yourself.
Nice one, jackass.

A similar ploy is used when it comes to foreign aid. Somewhere over yonder, there are brown people who are sick, dying, hungry, thirsty. But don't fret! For only three dollars a week, you could keep one of them fed, saving their life in the process. *So? What say you?* Well, if what we're told is true, their fate(s) are in my/your incredibly powerful hands. This, my friends, fellow consumers, countrymen, is the beauty of the free market – you get to vote with your dollar. Everything's up to you, from which of the grocery store's forty-seven types of peanut butter you want to boycott, all the way down to the quantity (and shade of brown) of 'unfortunates' you want to save. Once their souls have been put on the market and we're made aware, every cent we *don't* send their way is tantamount to an emperor's thumbs down manslaughter, or gross negligence at the very least – either way, morally indefensible.

Our limited geopolitical understanding doesn't help. We can't *really* wrap our heads around the thought that a middle-class person in, say, Bangladesh, might have an equal or better chance of a happy life than a poor person in Kentucky. Of course, it's an erroneous idea that there's a clear delineation between the haves and have-nots, and that any such list would be neatly organized by geographical borders. For most of us, though, irrespective of our political views, this misconception is pretty much baked in. Some places are there to be visited and pitied, while others aren't. Some places (and races) need help, while others are meant to do the helping. Passive versus active, voyeur versus viewed; you're one or the other, but not both. Just think, for American poverty-porn connoisseurs, it would be patently absurd to see a gaggle of teenage Rwandan volunteers getting bused into, say, inner-city Detroit, or Skid Row in Los Angeles. That's just not how things work. *Just what are they implying, those Rwandans? That we can't figure out how to do things on our own turf?* We tour *your* slums, motherfuckers, not the other way around.

Our conceptions of things like 'poverty' and 'privation' muddy things further. We, in our strange little Western bubble, have cranked our hedonic treadmill up to maximum speed and, as a result, our understanding of what life 'ought' to look like has been freakishly distorted. From up in our crow's nest, poverty becomes a binary distinction – if you're not up here with us, you're far, far below. From an angle this steep, everything down there looks pretty much the same (to wit: horrific). For the compassionate, this only engenders pessimism about the state of the world.

And it's in this malaise of guilt, ignorance, hubris, and self-loathing that industries like self-help – of which voluntourism is effectively an offshoot – can step in and make a killing. It's like the Catholic Church selling indulgences.

It seems like our challenge, then, when it comes to how we see the global South, is one of acceptance. Acceptance that we are lucky, and that we exist in a world infinitely more comfortable and more hospitable to human life than it's ever been. In a world that may not yet have achieved equilibrium, but one where conditions, when examined on a large enough timescale, do tend to improve. We assume that our steps forward have been uniform, instantaneous, and simple, when they haven't. What we've seen with the internet and technology is a total aberration, in the past it has taken centuries – if not *millennia* –for upgrades to the quality of human life to catch on.

Even here, though, I can feel my Samaritan training kicking in. It's all very easy for me to sit in my ivory tower and pontificate. *Everyone just needs to settle down. Historically, we've needed to throw a few million lives at our global problems before finding a solution. This might just be what we're watching in real time.* But is this not just a roundabout way of justifying the status quo? It feels too convenient. Is that really all it takes to get us off the hook?

Or is advocating for 'patience' *not* just a spineless apology, and in fact a perfectly reasonable stance? After all, for almost all of human history, keeping to yourself and letting far-flung geopolitical issues sort themselves out was your only option. Pretending otherwise would have had you burned at the stake for being insane, or a false prophet.

So, what were my options, then? I could try the humanitarian route again, albeit with less ignorance this time. But I wanted a bit of space from that crowd – and they probably wanted some space from me, too. I just couldn't hear one more Valley girl talk about how she'd become 're-entwined with the Spirit of Africa' after getting her picture taken holding a suspiciously drowsy-looking lion cub. (Such photos, I had come to learn, never captured the true magic of the moment. They invariably involved a couple of litres of sedative and the poor animal's handler, just out of frame, armed with a tranquilizer gun.)

In a perfect world, I'd visit South Africa as I would any 'regular' country. But this would require suppressing my superhero training. Was I up to the task? I hemmed and hawed for a while, but in the end the decision was made for me. At the beginning of my senior year, the head of the university's study-abroad office retired, and with the new head of department there was a rule change: transfer students like me, who'd previously been ineligible, could now apply.

꿈

With Trump in office, I knew I'd be putting myself in the line of fire – it didn't help that I look like someone who might've helped manage his campaign, or did coke with his sons. In my experience, the way people treated Americans abroad correlated with how much respect they had for the president. This was irrespective of what was going on in their own countries. I remember getting grilled by snobby English relatives about the invasion of Iraq back when I was seven or eight; never mind that Tony Blair had dutifully followed Bush right in.

Tensions had cooled during the Obama years, but those days were over. I could no longer ride on the coat-tails of my – well, my *parents'* – anti-racism vote (I'd been too young to vote in 2008 and 2012). There was nowhere to hide. Even our Alamo, blaming the electoral college, had fallen. Pennsylvania had gone Republican for the first time in several decades, so I couldn't even deflect blame. *Hey, we did our job, take it up with Ohio!*

Thankfully, all this was moot. It turned out that the only school near Cape Town that was accepting international students for the upcoming semester was in an area sympathetic to America's new president: Stellenbosch, whose town and surrounding wine region fancied itself as a bastion of tradition. (For reference, it wasn't far from here that we'd witnessed the scarecrow in blackface.)

ACT II

(FEES MUST FALL)

RED TAPE

The arrangements for my second stint would prove far more protracted and aggravating than those for the first. My university had a shortlist of affiliate schools it preferred students to visit, none of which were in South Africa, and it was reluctant to let me explore other options. It took months of negotiating with the administrators before I got their assent. Even then, it was only given begrudgingly; I had to sign a waiver acknowledging that I was organizing this trip at my own risk, and with my own funds. I managed to mediate a tenuous accord between the schools: I would first drop out of my US school before applying to the South African one as an independent student. The South African university had a specific classification for nomads like me, which I liked: *Freemover*. After the semester, I'd withdraw and reapply to the original university in Florida, who would plug in my South African credits and, hopefully, allow me to graduate.

But there was a snag. Two, in fact, in the form of basic credits that I was apparently missing. One was a paltry little half-credit course mandatory for all incoming freshmen. It gave you a brief orientation of the town, the campus, and general help with transitioning into adult life – living away from home, managing your time, etc. By the time the existence of such a class had been brought to my attention, however, I'd already lived alone in the town, hundreds of miles from Lancaster, for a year, making it redundant. The advisor of this course was, it turned out, my professor for international law. The registrar told me I just needed her signature and the credit could be waived. Given the nature of her field, requiring a grasp of things like logic, reason, etc., I was expecting our meeting to be a quick and successful one. I was *half* right: I was back out in the hallway in under a minute.

'This isn't like McDonald's,' she'd said with a frown, 'you can't just pick and choose whatever you want.' Ironic, I thought, coming from the industry putting tens of millions of young Americans in, on average, fifty thousand dollars of debt, under the pretence of letting them do exactly that: pick whatever (futures) they wanted. Everywhere else, the customer was always right. Not here, apparently. Luckily, the one other professor who had the power to waive the credit, a sweet lady in the archeology department, didn't even let me finish my spiel before snatching the paper from my hands and signing it. *Maybe this was a McDonald's, after all.* One redundant requirement waived, one to go.

When the registrar's office had perused my records, they'd found only an *advanced* algebra credit (which I'd transferred in from my first university). I was missing its 'basic' prerequisite. This meant that, instead of graduating upon completion of my semester in South Africa, I'd need to come all the way back to Florida and pay a full semester's worth of tuition (and rent) for a freshman-level class that met once a week. I spent several hours arguing that, *surely*, having passed the more advanced course made the basic one redundant. This was to no avail, so I offered a compromise: I'd take an equivalent class at my old community college over the summer before I left for my exchange. They agreed, with the stipulation that I had to get an A. My academic future (and thousands of dollars) relied on passing this course with flying colours – I couldn't leave this to chance. Fortunately, the instructor, a rickety old Chinese man, couldn't see anything happening more than a foot beyond his plastic podium; I sat in the back row and used a calculator app on my phone to complete every single assignment. (I made sure to get a few questions wrong here and there so as not to arouse suspicion.) *Credit secured.*

All I needed now was a visa. This was a new frontier. On my first trip, I'd stayed just ninety days, the maximum you were allowed to visit as a tourist. I'd heard rumours you could reset this imaginary deportation timer by taking an overnight bus north and briefly crossing into Namibia, but that option was off the table, as enrolling at Stellenbosch required an official permit.

The South African embassy's web page had an eighteen-point list of forms and other documents needed for the application. On the surface, these appeared elementary, but this checklist would prove to be as exacting as anything Hans Zimmer had curated.

All right, let's see here ... First on the docket was the vague request for applicants to show 'satisfactory proof of finances'. Because their website looked like it hadn't been touched since the turn of the millennium, it seemed wise to do a quick check-in with someone at the office in case there had been changes or updates to these requirements. When I called the handful of consulates around the country, however, each individual gave their own interpretation of the rule. Some implied it was a matter of demonstrating fiscal responsibility, and thus suggested I'd need to provide one- or two-years' worth of healthy bank statements. Others, meanwhile, implied that their visa agents would be swayed only by whatever impressive sum of wealth you could prove you had on the morning of your appointment. When they refused to put a number on what might qualify as 'satisfactory', I suggested that maybe a specific minimum wasn't worth stressing about. After all, the country's low cost of living and low exchange rate meant that even the meagre reserves of a college student would be enough to last a few months. Unfortunately, they didn't welcome my contributions, taking them as condescension typical of my ilk. They doubled down, emphasizing that there *was* a firm benchmark I would need to exceed. Not that the secretary in Washington planned to elucidate, of course. *'Was there anything else you needed, or would that be all?'*

The next item on the docket was to provide evidence of a clean criminal record. Fine. You can walk into any police station or any number of local municipal agencies and pay a few dollars for them to take your fingerprints and run a check on you. If you're seeking a slightly higher-stakes career involving international travel or government work (or, apparently, if you're hoping to study abroad), you may have to do a background check of the *federal* variety. This process is mostly the same, but costs more and means your fingerprints are run through the FBI's own, wider, database. The FBI's website told me that, for my convenience, they worked with hundreds of local fingerprinting stations across the country to provide these checks, and the results would be returned via an encrypted online document. I figured this would be fine, but decided to check back in with my friends at the embassy. Thankfully, the secretary confirmed my theory.

'Whatever the site says is accurate. If it says we need an original copy of your background check, then that's what we need.'

'Sweet. Just wanted to make sure.'

As a gesture of goodwill, I went above and beyond, getting both a state and federal check. I got my fingerprints taken at those two separate places, where I mentioned that I'd need an official paper copy of my results. They seemed confused by my request.

'That's not something any customer has ever needed before. In fact, I'm not even sure if anyone can legally ask for that.'

'But if they did?'

'I would just print off a copy of the encrypted file the FBI sends you.'

'OK, thanks.'

I whizzed through the rest of the checklist: I got a couple of passport photos taken and printed, got copies of my passport and birth certificate notarized, received a physical exam from my doctor, bought travel insurance, and filled out a pile of other forms.

I now notice on the embassy's website a memorandum announcing that they no longer allowed applicants to make reservations: first come, first served. The visa officer would be on-site from Monday to Thursday from eight to half past eleven in the morning, with no indication whether he would return in the afternoon. As usual, I called ahead to verify, adding that I'd be driving several hours to get there and, in the event of circumstances outside my control, would very much appreciate some flexibility were I to arrive a few minutes late. The receptionist had no time for this special pleading. She curtly restated the hours, adding that, even if, hypothetically, I was next in line at half past eleven, I'd need to come back the next day. They're very busy with extremely important work in the afternoon, you see; if you want to see the visa man, you ought to arrive early. This meant the next day's drive to Washington, starting before dawn, carried incredibly high stakes. I'd be shocked if homeland security didn't flag me down; my reckless weaving toward the nation's capital must have looked suspicious. Thankfully, I made good time and could drop back under the speed limit a few miles outside the Capital Beltway. With just under an hour before the agent clocked out for the day, I found parking near Embassy Row and hurried toward the South Africans, sprinting past the bronze, three-metre- (ten-foot) tall Mandela outside before bursting into their office out of breath.

Strangely, none of the urgent productivity and ruthless efficiency I'd been warned of by the receptionist was evident. This was a sleepy waiting room, with a portly woman behind a glass window asking visitors to sign in on a clipboard. As I did so, I noticed that only two other people had stopped by that day. There was only one small side office, presumably where the visa interviews were being held. One other young man was in the waiting room with me and, as he recounted the saga of his preceding days, any self-pity I felt after my three-hour drive that morning disappeared.

He was from Nebraska and, as he couldn't afford a plane ticket, had driven fourteen hours through the night. But his journey had actually started two days earlier. He'd arrived yesterday morning, only to learn that the three months' worth of financial records he'd brought were insufficient. The agent had decided that he'd need proof of having a few thousand dollars in savings over the past *six* months. So he'd run out to his car to fetch his laptop, which he used to show the visa officer some earlier records. But this wouldn't do, either.

'No, no, no. These must be printed and notarized,' the officer had said. The young man quickly did an online search for the nearest notary, a ten-minute drive away. The agent pointed at the clock, which had just gone eleven.

'You won't get back here in time – you'll have to try again tomorrow.'

With no other choice, the young man slept in his car around the corner. By the time I arrived, he'd already been turned away *again*, having needed to run to the printing shop to get a faxed letter from his parents stating they would support him in a financial emergency. Luckily, he'd made it back in time. As he got called in, I wished him luck. It worked: at the third time of asking, they accepted his documents. He emerged, looking more relieved than excited. Whether he'd genuinely met the requirements, or the visa officer had grown bored of toying with him, we'd never know. He wished me the best, and left to start his long drive home. A bark came from inside the office.

'Next!'

I let myself in and sat down in front of the window. I confidently placed the stack of documents in the aluminium tray under the plexiglass, hoping the thickness of my portfolio gave it (and me) an air of legitimacy. Alas, like a police bloodhound trained to smell blood traces even after visible evidence has been scrubbed away, the wiry little South African coloured man immediately found a weakness. Without looking at any of my other documents, he took out a page from somewhere in the middle of the stack. *My background check.* He looked at me for the first time since I'd sat down, rubbing the corner of the paper between his thumb and index finger.

'Wrong paper, my friend.'

'But my results are totally clean. I can even show you them on the FBI's website.'

'Doesn't matter. It must be printed on the correct paper.'

'All right, fine. But can you at least check the rest of my documents to make sure they're acceptable? That way we can at least make some progress while I'm here.'

'Not possible. I can only evaluate an application in its entirety.'

I slunk back to my car, thwarted. As I got in, I realized that, in my shell-shocked state, I'd forgotten to probe for more information. *What did he mean, 'wrong paper'?* I glanced at my watch, and my stomach sank. *11.31. You're on your own, kid.*

The plot thickens once you get home and start calling around to these official FBI background check agencies to ask whether they offer, or are even aware of any 'special' paper. The consensus is *no*. Hours of trawling through the internet also fails to provide evidence that any such stationery exists. Apparently, no other prospective visitor has ever had this problem – not even the Nebraskan. The farce continues when you call any of the other South African consulates in the country to inform them of your findings and ask if you can submit your background check to them as the original, online document – a method, you mention, that the FBI *itself* recommends. No, you learn: the *only* acceptable rap sheet is a rap sheet printed on a mysterious, obsolete paper stock.

The receptionist in Washington does not seem to appreciate your suggestion that the only explanation for all this is the embassy having a vested interest in pushing this stock, perhaps earning some sort of commission for every client it can secure. The woman implies that I'm not treating their institution with sufficient respect.

'I apologize, ma'am. Could you please tell me where it's possible to find a background check provider that will give me the results on your special paper?'

'We don't have that information.'

Her ineptitude/indifference is echoed by the consulates in New York and Chicago. But at least they pick up; the phone line to Los Angeles rings for several minutes before cutting to a dial tone. I give up and go back to canvassing every background check agency within about a two-hundred-mile radius. After a few hours, I find one that might work.

'We aren't really supposed to print on that stuff anymore – the FBI says it's way less secure. But let's see what we can do.'

After an hour's drive, I arrive at their address, an outbuilding in an industrial park. A guy taking a smoke break stands beside the entrance in front of one of the disabled parking spots. He wears a leather jacket and basketball shoes and has dark hair gelled into a Mohawk.

'We're cash only.'

'No problem.' I'd assumed as much, and had stopped at a gas station ATM on the way here. He hurriedly drops and stomps out his cigarette before buzzing me in, then leads me into a dingy, one-room office where faux-wood panelling covers the walls and old, stained ceiling tiles hang above a peeling linoleum floor. He takes my fingerprints at the check-out counter. I'm back in the car after just a few minutes and I'd now spent almost three hundred dollars, but, by the grace of God, the special, oily paper arrives in the mail in a couple of days.

On my next visit to the embassy, the man behind the bulletproof glass (his need for which I now understand) takes the stack of papers I hand him and flicks through them for several seconds, uninterested. He then flips back to the beginning and slaps a staple into the top left corner before reaching for his rubber stamp, which he presses first into the ink pad then briskly onto the first page. All of this is done with theatrical panache, an affect that continues as he spins his chair to the small filing cabinet behind him and puts my application into one of the drawers.

'Five to ten business days,' he says, pointing to the door behind me. In shock, I stand to leave. I thank him, but it seems he's already irritated by my presence. He mumbles a response without looking up from the other assorted papers on his desk, which he's already begun to shuffle through. As soon as I've pulled the door ajar, he summons another victim.

'Next!'

I get back to my car thrilled, albeit slightly peeved. After all this trouble, *that's* how little any of this actually meant? *At least let me know you're in on the joke, for God's sake.* Oh, well. Never mind that now. We're South Africa-bound, baby.

FEES MUST FALL

Stellenbosch is about an hour's drive inland from Cape Town, nestled amid the surrounding Hottentot mountains. These shield the town from the offshore winds, keeping it much warmer than the peninsula in the summer months. Though less precipitous than their cousins on the Cape, these mountains are no less captivating. At dusk, their sandstone glows a rich, lavender-tinted sienna, as if releasing the heat collected during the day. Beyond lending itself to a charming lethargy, the topography is ideal for viticulture; any arable land that isn't too steep to farm is covered by rolling vineyards. The historic little town relies heavily on the wine industry, but also on the university, which creates jobs and attracts thousands of students from all over the world.

I was placed along with the other international students in a cluster of small dormitory buildings on the edge of campus. Of my three room-mates, two were American, as were three of the four girls who lived across the hall. In fact, most of the other students in our building were from the States. I learned that there were four or five major study-abroad companies that bundled these students together and shipped them overseas. By the sound of it, there were some benefits to travelling through this kind of outfit. For instance, they sent a designated liaison officer along with your group, and it sounded like the overall process was very streamlined. You just submitted a short online application, handed over your usual tuition fees, then the travel company took their cut and helped you get your visa and enrol in your exchange school. But all this convenience came at a price (other than not getting to explore the arcane world of discontinued paper stock, that is). The students who'd ceded control to these agencies were still paying the exorbitant tuition fees of their home schools: up to thirty thousand dollars for the semester. For as much hassle as the process had been, by making it to Stellenbosch independently, I paid only about two thousand dollars.

While the Americans seemed unvexed by our extortionate education system, the South Africans had reached a boiling point with theirs. Tuition costs, albeit affordable for international students like us, were prohibitive for millions of the country's own young adults – especially so for those of colour. Over the past couple of years, nationwide protests had been held in cities and on campuses under the broad thematic umbrella of 'Fees Must Fall'. Initially, demands were for reduced fees across the board and increased bursaries for those in need. Soon, though, their scope increased, and there was a growing clamour for schools to disavow themselves from their colonial past.

The trouble was that the country had only been desegregated for a sliver of its modern history. Most of its universities and public infrastructure had been founded and sponsored for decades by people who had a vested interest in maintaining imperial supremacy. Today, of course, certain practices are easily identified and redressed – slap together some new world maps with Africa in the centre, instead of the United States or the United Kingdom, postpone all pencil tests and eugenics experiments indefinitely, take down the 'whites only' signs on public bathrooms and water fountains, etc. But once all the low-hanging fruit are picked, things get a little bit trickier.

A topical example was the controversy surrounding the Rhodes Scholarship, an endowment of immense prestige which has helped thousands of students, including non-whites, become influential in the sciences, public service, and the arts. Of course, the programme's namesake, Cecil John Rhodes, also dabbled in other passion projects – namely, the advancement of colonial interests, as he administered South Africa while it was under British rule in the late 1800s. In fairness, he had a real knack for the job, even managing to get an entire country named after himself – Rhodesia, now Zimbabwe.

These issues would be far simpler were it not for subjective human experience, because nothing happens in a cultural vacuum unaffected by social trends. Debates often become proxies for larger social concerns, such as the O.J. Simpson trial and America's racial tensions. When factors that might have been considered damning in one case are ignored in another, reaching a consensus as to how or even *whether* we should separate the art from the artist is nigh on impossible. A perfect example was the debate that had raged throughout my childhood: what should we do about Michael Jackson? Everyone could see the allegations piling up, but it was impossible not to remain spellbound by his prodigious talent. *Thriller*, it turns out, continues to be catchy, at least for the millions of people who've bought – and still buy – the album; and that's not even considering all of Jackson's humanitarian efforts. It was the same with cyclist Lance Armstrong who, for all his misdeeds, had inspired millions of people and made immense contributions to cancer awareness and research.

There could be some fascinating philosophical questions to ponder here. We might, for example, consider the devastating effects of unresolved childhood trauma, or even question the broader systems that create these sort of polarizing characters. Unfortunately, we have neither the time nor the inclination for these sorts of frivolous pursuits.

I can't help but wonder if our struggle to reconcile aspects of these complex – but nevertheless real, and human – characters is due, at least in part, to the training we receive as kids. We are bombarded with superhero movies and cartoons that give us unrealistic expectations about how things ought to work. The good guys/underdogs always win, and the bad guys are always taught the error of their ways. *Happily ever after; roll credits.* Unsurprisingly, we then try to apply this framework to the real world. From our point of view, it's unthinkable that we won't help catch and vanquish the bad guys, whomever they may be – let alone that we might not be the good guys at all.

❧

For now, though, it seemed that Rhodes and his contemporaries had not yet been disgraced. Around Stellenbosch, at least. Streets, dormitories, and various other school buildings and facilities remained dedicated to: Smuts, Coetzee, du Toit, du Plessis, le Roux, or the van-*somethings*: Zyl, Dyk, der Westhuizen.

The similar names of the Afrikaner professors and students was due to the small size of the original Dutch colony, so only a few of these family names had ever existed on the continent. Bad news for their gene pool, maybe, but this could potentially make for a useful alibi: if pressed, they could easily have claimed that their shrines, tattoos, or plaques were for more obscure, less controversial figures; a classic sleight of hand. *Oh, the hospital? It's actually named for a Kevin Mengele – no relation. He was an alumnus of our medical school and was very generous over the years. The sign only had room for his surname.* Of course, my clever little plan might not have worked for the more historically accurate statues and busts. Indeed, the large bronze cast of Rhodes in front of the University of Cape Town had recently been torn down after years of heated debate, and many others like it around the country, and in the UK, were next on the docket.

There had been similar incidents in the US. I wasn't totally up to speed on South Africa's dignitaries of yore, but I assumed the gist was the same: invariably, it comes to light that a politician or military hero actually ran a plantation with thousands of slaves, and was known for being a violent drunk, an adulterer, and a thief. Following these revelations, it's suggested that perhaps there *shouldn't* be a massive statue of him in the town centre. It's at this point that the country's self-appointed heritage-protectors and devil's advocates come out of the woodwork to inform us that the poor old fellow was just a 'product of his time' and we therefore oughtn't hold him accountable to today's sensibilities. The logic of this argument becomes a little more tenuous when these history buffs claim that those dopey slave owners would also have had the incredible foresight to *conceive* of heavy automatic assault rifles and the wisdom to completely support the US citizen's right to drive around with one in their truck.

I'd long since lost faith that there was a coherent narrative in the US. I had seen more Confederate-flag bumper stickers *north* of the Mason-Dixon than in the supposedly more conservative Florida. It seemed that these flags had become more of a team logo, a signal to one's ideological in-group. This was by no means a partisan phenomenon: with no more than a couple of data points, you could pretty much bet your bottom dollar on how someone voted. But this is simply the logical end point of a first-past-the-post system. So dominant is the two-party duopoly that its constituent blocs are forced to appear increasingly polarized, creating a false dichotomy. Contrived as the 'rift' may be, as it widens, the social stakes rise and there becomes no greater shame than to be suspected of sympathizing with the other team.

One man, the father of a girl I had dated in high school, wanting to simultaneously prove his patriotism and avoid aiding John Kerry's 2004 campaign against George W. Bush at any cost, swore to never again consume a drop of Heinz ketchup, or any of the brand's other products. If you're having trouble connecting the dots on this pledge of fealty, Kerry had been married to an heir of the condiment magnate. Although almost ten years had passed since that election by the time I met the man in question, he was still shipping in George Washington-themed, W-brand ketchup from a small outfit across the country at twenty dollars a bottle (plus shipping and handling). I wonder how he handled the news that Donald Trump was actually related to the Heinz family, too. Did he suspend the boycott for a few years? Or was he in too deep, and now had to pretend that he actually preferred the flavour of his boutique brand?

❧

By contrast, South Africa's system of proportional representation interested me, as it seemed to foster a more varied range of ideologies than in the US, where fringe parties and voices were so easily shouted down. Occasionally, we see glimmers that other options may have existed, but these are never addressed in good faith. For example, communism, we're taught, has accounted for over a hundred million deaths. Capitalism, meanwhile, hasn't accounted for *any*, thank God. The most left-leaning party in the United States is, in effect, the Democratic Party, which would be considered centre- or right-leaning in most other developed nations.

Here in South Africa, plurality was possible, and there were multiple parties with genuine clout, each with their own distinct objectives, including things like economic reform, land restitution, and Pan-African unity. Most of these parties agreed that South Africa needed to liberate itself from the yoke of the colonial powers. This sentiment was not exclusive to the left. Stellenbosch, along with many other conservative, heavily Afrikaans, regions in the country, had its own qualms about imperial omnipresence. The town took issue with the country's educational institutions and economy now conducting operations mostly in English, which led to the diminished presence and influence of their own language and culture.

Based on my experiences in the handful of bars and clubs scattered around campus, however, which also had Afrikaner names, their language appeared and sounded alive and well. It's based on Dutch, but has undergone enough linguistic divergence after a couple of hundred years that it's now officially a distinct language. I speak neither language, but this did nothing to stop me getting harangued on various occasions, by drunk Willems, Pieters, and Johans. Afrikaans culture, and their tongue in particular, was under siege, you see. And not just from the blacks, either. By *everyone*.

This was made particularly clear to me at a nightclub called Catwalk, which was favoured by the other international students, but I only suffered through a couple of visits before I swore to never return. The stripper poles were downstairs, but the smoking section was upstairs, and this took precedence.

Funnily enough, none of the Americans smoked – it seemed that for all our self-destructive addictions to sugar, processed foods, and lobotomizing media, we'd somehow done a good job at warding young people off tobacco. But the habit was hugely popular among other students, especially the Europeans, who took full advantage of cigarettes costing a tenth of what they did back home. The catch was that the second floor, through which you had to venture to reach the bar, was reserved strictly for a genre of Boer folk rock and dance called *sokkie*. Chilling enough, but it was made even more so by the room's lighting and colour scheme. Lasers and strobe lights roamed over eerie phosphorescent symbols and occult hieroglyphs painted at random angles and sizes around the pitch-black walls and ceiling. This crippled all depth perception, making the exits and even the very boundaries of the room impossible to discern. It was as if we were in some sort of cosmic purgatory, void of time and meaning. Fun if you're on hallucinogens, maybe, but *not* so fun if you're trying to escape – or even just looking for a beer. It was here that the rambunctious young Afrikaans confronted me. Given that participation in their little dance seemed mandatory, I figured that this was their grievance. (I would have joined in, but it looked like a cross between an organized square dance and a traditional big-band swing performance – I had no prior experience in either.) I tried the usual peacemaking techniques, namely flattery, complimenting their camo dungarees or farming wellies, to no avail.

Fortunately, one of these incidents occurred while I was standing with a fellow exchange student who spoke Dutch and could therefore give a rough translation. Apparently, I looked or sounded to these local guys like an Englishman. I had a feeling it wasn't the time to explain to them that, actually, *mein vriend*, I was only *half* English, and had in fact grown up in the US.

But I learned that even this would not have sufficed, because for many Afrikaners, *any* outsiders, even *white* ones, are suspect. Thus, even Aryan-looking little me represented a clear and present danger. The whole Boer War fiasco hasn't been forgotten, it turns out; the 'English' are still in their bad books.

Coincidentally, similar terminology was used back in Pennsylvania, where the Amish used the term English to refer to anyone outside their community. In fact, the groups seemed kindred spirits: both were small, Dutch or Germanic farming colonies who, despite retaining little-to-no loyalty to their countries of origin, still clung to a unique, if slightly crude, variant of their mother tongue. Kind of endearing, if you discount the white nationalism and penchant for inbreeding. In any case, I now had a better grasp on their distress: if they didn't like the look of *me*, there was no chance they trusted *anyone*, and this must have been exhausting.

A couple of times, it sounded like they were calling me 'soupy'. This didn't make much sense, until I learned they were actually saying 'soutie'. Granted, this didn't make much sense either. *Soutie*, it turns out, has nothing to do with soup (or soot). It's a slangy version of the term *soutpiel*, which means *salty (sout) penis (piel)*. In the colonial days, non-Afrikaans whites weren't seen as *true* South Africans, as they had one (figurative) leg down here and one back in Britain, which therefore left their willies hanging in no-man's-land: the notably salty Atlantic Ocean. This may have been intended as a pejorative but I took it as a positive, as I had done with *Preocupado*. I'd finally found a word for how I'd felt as a kid, alien to both the United States and the United Kingdom. With my discovery of South Africa, I'd added a third point of contact, and completed the triangle of trade, but still had nowhere dry to rest my [pruned] johnson.

৯

If I held on to a sliver of hope that the farmers' disdain was rooted in some sort of Pan-African, anti-imperialist solidarity, this was extinguished a week later when I saw a small group of black students get turned away from the bar for which my friends and I were also in line. While the bouncer checked our IDs, one of the girls asked him why the other group hadn't been allowed inside.

'We didn't want them to get into any trouble.'

Well, I thought, *at least I had the privilege of being trapped in the epileptic-fit-zone while getting spittle sprayed into my ear by a bleating skinhead – those below me on the pecking order weren't even allowed in!* I couldn't help but recognize the bouncer's phrasing as the sort of sinister passive voice a couple of good-fellas might use to talk about 'what a shame it would be' if the owner of the corner store they were leaning on was found 'sleeping with the fishes'. It was as if he was actually looking out for the group he'd turned away.

The incident was made slightly more confusing, though, given that the bouncer himself was also black. Not that he seemed bothered, the directive had clearly been passed down from above; he was just following orders. After a few of our probing questions, he explained his employer had wanted to increase security in response to the crime wave the town had suffered over the last few months. For as peaceful as Stellenbosch seemed, it was not without its share of gruesome tragedy. Mirroring my arrival in Muizenberg a few years earlier, when I'd been notified that there had been a break-in and stabbing a few days earlier, a revelation had come to light immediately after arriving on campus.

Just a couple of weeks earlier, the body of a female Afrikaner student had been found on a farm nearby. After meeting her friend at a bar in town (across the street from Catwalk, as it happens), she'd insisted on giving him a ride back to his apartment just beyond campus, on a road bordering the nearby township. As she parked her car, a group of coloured men had carjacked and kidnapped them, beating her friend nearly to death before raping and killing her. During their trial, the bowling-ball-sized stone the killers used to crush her head had to be carried into the courtroom by two large security guards. The men were convicted, with the judge noting the almost total lack of remorse exhibited by the killers (members of one of the country's biggest and most dangerous African nationalist gangs) throughout the trial. As you might imagine, this ordeal did little to allay the Afrikaners' concerns.

Unsurprisingly, distress over the waning influence of Afrikaans garnered little sympathy from the black population that mainly felt the language of apartheid *ought* to be relegated. But the question, as in any political deposition, is what to replace it with – English, after all, had been used for the same ends. Matters are complicated not just by the country's multilingualism itself, but because it is not delineated perfectly along racial lines: most of the coloured population speaks Afrikaans as a first language, for example, but a large percentage also speaks English. Among the black population, meanwhile, the confluence of tribal and ethnic groups means there are more than two dozen unique languages and dialects. We think of South Africa's sociopolitical situation as being a black-and-white issue – literally – with the country's black majority uniting to overthrow the white minority. But this is an oversimplified, Hollywood-ized narrative that was more a construction by and for the West than anything else. Bitter tribal rivalries go back centuries, long before apartheid – even the ANC itself is made up of multiple factions.

In the wake of decades of cultural erasure, the post-apartheid democratic government offered linguistic representation through a revised national anthem, ingeniously incorporating the country's five most-used languages. As a result, the song includes clicks, glottal stops, and other guttural sounds. A stroke of passive-aggressive genius, I thought. What better retribution: before their rugby starts, force the white folks to sing in tribal languages that are almost impossible for non-native speakers to replicate.

By the time I saw the anthem attempted in person, though, the crowd had mastered it, and their rendition was magnificent. A classmate had offered me a spare ticket to see the national rugby team, the Springboks, who were hosting New Zealand's world-famous All Blacks; this was an opportunity I couldn't pass up.

NEWLANDS/OLD HABITS

The visiting team (whose All Blacks name refers to their iconic uniform) was a perennial powerhouse and a global brand in its own right, but the rivalry between the teams had historical significance of its own. The Springboks had beaten the All Blacks to win the Rugby World Cup in 1995, a tournament hosted by the recently democratized South Africa, with its new president, Nelson Mandela, watching on. At that point, the Springboks team was still almost totally white, so it was fortunate the new anthem had not yet been ratified; the players and supporters still had a few more years to learn the Zulu and Xhosa parts by heart before needing to sing them on live television. This was just as well, as the Kiwis had their own pre-match display of fraternity, which would likely have stolen the show anyway – the *Haka*, an ancient Māori war dance.

On the day of my visit, the stadium, packed with fifty thousand rowdy fans, fell into a reverential silence as the All Blacks assembled at midfield in triangular formation. As they began, we watched on, transfixed. Although my friend and I were seated at the far end of the stands, we could hear their chants and cries, their appeals to the heavens for good fortune, and even their rhythmic, thunderous stomps – this wasn't too surprising since their thighs were thicker than my torso. The home team stood a few metres away, arm in arm, trying to look as unperturbed as possible as they held the gaze of the screaming men before them. Although the power dynamic may seem unbalanced, as only one team gets a sanctioned opportunity to frighten the other, the performance is considered a gesture of respect; it was clear that both groups relished their roles.

The dance had fascinated me since I was five or six, when my dad had visited New Zealand and returned with a little booklet for me, which I'd studied and carried around for weeks. Its dozen or so pages explained the historical and cultural context of the dance, as well as its choreography, which included lots of synchronized stomping, chanting, and grotesque, gargoyled faces from the participants, all while staring menacingly into the eyes of their opposition. As a kid in Amish country, I'd found these mysterious, hypertrophic men with their face tattoos and piercings impossibly exotic and terrifying, but also a little ridiculous. Men didn't dance or sing together. Certainly not big, strong ones who wanted to be taken seriously. I'd seen our athletes sing the national anthem, or even celebrate scoring a touchdown or home run with a silly dance they'd planned with a few of their teammates, but this was different. It was far more visceral, far more intimate. The idea that choreographed theatrics, vulnerability, or self-expression could be respectable, and even *intimidating*, was *not* what we were taught back in farm country. *Why mess around with all that nonsense when you can just send a couple of fighter jets over the stadium?* But these All Blacks were proof that our conceptions of masculinity and its associated attributes – dominance, aggression, bravery – weren't as inexorable as we'd thought. When they concluded, a deafening roar erupted from the entire crowd: Springbok fans lauding their team's defiance, All Blacks fans claiming the ritual's success. The New Zealand team seemed remarkably well represented, as I'd noticed earlier, when we'd been wandering around, enjoying the pre-match ambience.

The stadium, the second-oldest of its kind in the world, was, not incidentally, situated in one of the posh, quiet suburbs outside Cape Town, tucked beneath the mountain alongside the university and surrounded by private prep schools. Here, where the eastern side of the mountains does not yet directly border the ocean (unlike Muizenberg and the other towns along the peninsula, you're more protected from the sun and wind and, luckily for me, their drying properties. Indeed, I noticed remarkably few clouds of nosebleed-inducing dirt floating around as I explored these neighbourhoods, which are markedly more lush and verdant than their counterparts in the poorer areas. On a match day for the rugby (or the cricket, which has its own ground next door), large beer gardens and marquees were set up in the wooded clearings between the stadium and the nearby nature trail that led to the national botanical garden and ran alongside a meandering stream. Vendors lined this path and the residential streets nearby, selling hot dogs, snacks, and memorabilia: jerseys, flags, and hats for both teams were on display. (Although I noticed that no vuvuzelas were available, or even audible.) I heard a few Kiwi accents, but most of the fans in All-Black gear were actually South African, and most of them were coloured or black. There were thousands of these fans, easily as many as for the Springboks. I assumed this was partly due to the hegemonic success of the All Blacks; it's not unusual for dominant teams to garner an international fan base. On a given weekend during the UK's Premier League soccer season, for example, South African bars were packed with locals in Manchester United, Arsenal, and Liverpool shirts. As I waited in line for a beer at half-time, I asked around and got a more plausible explanation.

It turned out that the tall, ginger fellow who'd been standing in front of me was something of an authority on the topic. Alasdair was his name; he was a dyed-in-the-wool Capetonian and a lifelong rabid sports fan who now worked for the very agency that managed the Springboks' official supporters club. He told me that Mandela had realized that hosting the 1995 World Cup was a unique opportunity for South Africa to show the world (and itself) it was healing, and so he made the controversial decision to keep the national team's name, logo, and colours unchanged, despite the country's governing body for rugby and the national team having been inextricably linked to the apartheid regime.

The official party line was that this would promote the message of unity, forgiveness, and national pride. For many, however, Mandela's motivations were slightly more cynical: he'd been well aware that altering the iconography would have alienated long-time fans on the domestic front, many of whom still held financial and political power.

For *other* fans, all this simply came too late: during the apartheid years, a huge percentage had switched their allegiance to the All Blacks, whose players, national government, and supporters were outspoken against the South African regime. Although the Springboks and other club teams throughout the country were now integrated, a lasting stigma around the sport was unavoidable: rugby remained a signifier of cultural conservatism. I'd seen this before; it wasn't too different from the college-football-obsessed university towns all over the South and the Midwest back in the US.

This sense of provincial stubbornness was especially palpable in a town like Stellenbosch that was a rugby hotbed with a sixteen-thousand-seat stadium of its own. Its geographic seclusion and historic infrastructure seemed to encourage a certain romantic traditionalism. In the soft haze of the valley, my feeling was that as long as the rugby team kept playing and the drinks kept flowing, momentum for drastic, structural change would be tricky to muster up. But it wasn't just the Afrikaners who longed for a bygone era; quite a few of the *visiting* students did, too.

COSA NOSTRA
(OU, PEUT-ÊTRE: 'NOTRE CHOSE')

A university from Paris had sent over two dozen undergraduates, who lived in the neighbouring dorm block. Figuring that they were my best chance of making non-American friends during my trip, I spent most of the first few weeks with them, joining them on day trips into the city and other local excursions. Despite them all claiming to have deliberately signed up for an English-speaking country for language immersion, it seemed the novelty was short-lived: with increasing regularity, they reverted to their native language. This was their prerogative, obviously, but it was particularly frustrating in situations where they'd agreed to spend time with us English speakers beforehand. One day, for example, I'd organized a beach trip with a few of the Frenchies and the American girls across the hall. The girls filled one of the two taxis we'd called, so I offered to ride with the French in the other. As I approached the car, one of the guys (whose English was almost fluent, as he had a mother from Australia) said, with sincerity, 'Jack, it would be best if you sit in front with the driver. We want to speak with each

251

other in *our* language back here.' The rest of the afternoon continued along this trajectory.

This was all very conflicting. *Could I really be critical of people speaking their own language?* What made things worse was that my frustration was not without a tinge of envy: even the worst English speaker among them was close to being bilingual. I knew only a little bit of French – *merci; bonjour; I surrender!* – you know, the basics. That said, nearly all of the French had attended private schools where English was required, and the remainder had spoken a second language intensively from quite a young age, an advantage most kids didn't have back home. Despite the United States having no official language, there exists a general disinterest, if not an outright disdain, for non-anglophone cultures. Sadly, the characterization of our evangelical megachurch pastor thumping a copy of the King James Bible while proclaiming that, 'If English was good enough for Jesus, it's good enough for our schoolchildren!' seemed to be the de facto stance of the government and the public education system. Self-interest, we reason, is the only universal language, and we speak *that* just fine.

I know I'm not going to rustle up much sympathy for any country whose language is the global lingua franca, but think of it this way: language is an invaluable tool for connecting us with other people. As such, sticking to English-only will just reinforce not just the stereotypes of Americans and English being brash, incurious travellers, but our own unhelpful attitudes about the outside world (namely, that it exists to accommodate us).

252

My own shortcomings notwithstanding, I just couldn't completely absolve the French, whose efforts to cooperate were so brazenly half-assed. I knew how often they feigned bewilderment to escape any conversations that bored them, or began talking about us while we were still in the room. The more polite ones would pull out their phones and start sending messages about us into their group chat (which they titled 'The French Mafia'). As it happened, the only topics in English that reliably held their interest were critiques of the US. They resorted to this after they learned (to their disappointment) that my room-mates and I had not voted for Trump. Hot-button topics included foreign policy. Confirming what I'd observed over the years in the UK, folks from the imperial core were only taught about *other* countries' colonial misdeeds. The French, for example, had no idea that their government still collected billions of euros a year from its former colonies in Africa.

As the designated representative of the Anglophone world, I also fielded criticism based on lazy cultural stereotypes. This was much more boring; I can only be reminded so many times that 'Americans are stupid and fat'. Didn't these Frenchies know the golden rule of travel? You're only allowed to criticize your own country. Admittedly, this applies only while abroad; on US soil, criticism is met with, 'If you hate it so much, pack your bags and leave.'

Obviously, not all generalizations are unwelcome. If they're complimentary, you're likely in the clear; we pick and choose what and who we, and our countries, ought to be associated with. I remembered Andy Murray, the tennis player, remarking how he was passed off as Scottish (his country of birth) by England's famously spiteful media whenever he lost, but claimed by them as British – still technically true, but conveniently, a much wider umbrella – whenever he won.

Things get particularly sticky when it comes to low-hanging fruit. An observation being empirically true is no guarantee that it will be well received – the *opposite*, if anything. It's only human to be less receptive to criticism than praise, but this can be exhausting to navigate, especially when petty defensiveness is disguised as moral righteousness. It's important, I think, to confront challenging ideas and examine our own beliefs. Now, I'm by no means saying that all opinions are worth engaging with …

One night, one of the Parisian guys brought up fashion, under the broader heading of 'Things France Does Better than America'. He pointed out that his people were miles ahead of the United States in this field and had set a precedent we could only crudely imitate, to their long-standing amusement. I pointed to his footwear.

'But you're wearing boat shoes.'

'Yes, designed in France. An iconic style.'

Designed in *where*, now? I could feel the red mist descending. *First they came for Elmo, and I said nothing … But now they came for our Sperrys.* I'd let that little student back in Muizenberg get away with defiling one of our cultural cornerstones, but that was different: he hadn't known any better. This *snail-guzzler*, on the other hand, knew exactly what he was doing. So this was no time for kumbaya, patchouli-oiled bullshit. I had too many years of experience to let this smug jackass get away with spreading fake news – I'd worked at dozens of country clubs and been to more than my fair share of disgusting frats and house parties over the years. Boat shoes, whatever you thought of them, were a wardrobe staple for any white dude above the poverty line. Any self-respecting backward-baseball-capped frat guy wouldn't leave the house without them, *ad absurdum*: I'd played basketball with one such guy just before I left for my exchange – he played for hours in Sperry Top-Siders with no socks. The tan leather was soaked dark with sweat, and the smell, as you can imagine, was putrid. I thought back to that American hero. For his sake, if nothing else, I had to set the record straight; this was stolen valour. *'Look here, you slimy little frogs. Don't forget, you'd be speaking German if it wasn't for us.'*

I wish. But I just played it cool:

'Just so you know, chief, boat shoes are American.'

Of course, he and the other Mafiosos were incredulous, and hurriedly whipped out their phones to consult the web. After a few minutes of panicked scrolling, they realized: *L'Américain was right.* With this, they fell into a disappointed sulk and the evening petered out, but that was fine by me. *You don't have to like us, but you will respect us.*

This was about as contentious as things got on the social scene, although there were a few notable exceptions.

PRESSURE BURSTS PIPES

Semester-long university exchanges proved to be emotionally taxing for everyone, particularly for those with partners back home. Any prolonged separation is difficult, of course, but there seemed to be an additional pressure on the travellers, who needed to live up to the expectations of such a 'life-changing' trip. This, in turn, precipitated an undercurrent of resentment for those who'd been left behind in the doldrums of mundane life. Social media ensured these poor folks had salt rubbed in the wound relentlessly. This led to a vicious cycle of jealousy and revenge. During the latter half of the semester, we started to see lots of visits from partners back home who could take it no longer.

Some couples seemed more suited to the emotional roller coaster than others. One of the French guys, for example, confessed to his girlfriend back in Paris that he'd kissed someone at a nightclub. She jumped on the next plane down to set him straight. She was in town for a week, during which he managed to get caught kissing yet another girl. The girlfriend *still* didn't seem to mind: they got married a few months later. (It's a cultural thing, apparently.) But there were other couples whose long-term success seemed less assured.

One night, as I sat with a few of the American girls for dinner, I couldn't noticing that the one next to me was ignoring her food, with her face illuminated by the phone screen she was furiously tapping beneath the table. We knew she'd been embroiled in relationship drama for months. Whenever she joined us on a hike, grocery run, or any other excursion, she'd need to send her boyfriend a picture of her surroundings every half hour or so. In fairness, he, too, would be sending a constant stream of pictures confirming his whereabouts. We only got a chance to observe this bizarre ritual sparingly, when she was able (or allowed) to join us. Her extracurricular schedule mostly comprised locking herself in her room and embarking on several-hour-long phone calls, which almost always turned into screaming matches. Well, from her end, at least, in the form of vehement denials that she'd been misbehaving, alternating with (even louder) reminders that he'd cheated on her just before she left for South Africa, but it didn't seem much of a stretch to assume that the venom was flowing both ways.

On the evening in question, the photographic-evidence demand was invoked as usual, but its intensity seemed to have been dialled up. She got up to order another drink at the bar, but forgot her phone. A new message arrived, and I peered at her screen: 'Where are you? Go to the bathroom right now. Take off your clothes. Send me a pic of your body.' I quickly picked up a menu and pretended to scan it as she returned. She read the message, then glanced around to see if anyone had seen it. Nobody else had. With the colour drained from her face, she slunk off to the bathroom, emerging a few minutes later looking sheepish and slightly dishevelled. It would have been sad, if it wasn't all so confusing. For some reason, the topic of infidelity came up as we ate, and she got into a heated argument with one of the other girls, claiming that it was *always* the fault of the 'homewrecker', rather than the partner themselves. The predictable stance of someone trying to rationalize their decision to stay with an unfaithful partner. Yet it was clear she'd by no means forgiven her boyfriend. On the nights she would come out for a drink, she'd say she wanted to get back at him for cheating on her, and would find one of the French guys to make out with to make herself feel better. Midway through the year, her boyfriend showed up unexpectedly. They acted like nothing was wrong; indeed, he was thrilled. He spent most of the week-long visit blackout drunk. Fortunately for him (but not for her room-mates), he was still able to perform sexually. Each night, the couple engaged in two- to three-hour-long lovemaking sessions that woke the whole apartment. Almost immediately after he left, the phone calls and shouting matches resumed.

Then there was the group of three students from a college in New York. The girl, a Latina, was attractive, at least in the eyes of the two large dreadlocked guys who devotedly followed her around. A couple of months in, her secret got out: she had a fiancé back home. This left her henchmen, their time and effort having been wasted, in a regrettable position. But when one door closes ...

Another girl, Megan, a Bible Belter, mousier and more unassuming than the Latina, had been regularly buying weed from one of the two fellows; happily for him, he was soon able to parlay their professional relationship into a romantic fling. After a couple of weeks, however, his client-cum-lover confessed to him that she had slept with someone else – the German girl who lived across the hall. Megan's transgression was forgiven, but it was only a week or so later that she found herself in another sticky situation – roughly *twice* as sticky, in fact.

She invited her boy-toy to pop over one day after class. When he knocked at the door, she pulled it open seductively to reveal she wore nothing but underwear, and beckoned him inside. She then realized, to her horror, that he'd brought his dreadlocked associate, who'd been standing outside the doorway. The dynamic duo quickly managed to convince the girl that there was no need to feel ashamed or, evidently, to even get dressed. She let them both in, whereupon 'one thing led to another'. The technical term for what eventually transpired was a 'spit roast'. (*TIA, baby!*)

This was all consensual, but we soon learned (news travelled quickly around the dorm block) that Megan's escapades over the last couple of weeks were not without a victim, namely Kevin, her fiancé back in Iowa, an aspiring youth pastor, with whom she'd just signed over a deposit for an apartment that they planned to share upon her return. One her room-mates convinced her that she had to tell him what she'd done, and she reluctantly scheduled a video-call confessional for the next day. *Uh-oh.* Surely her death warrant had been signed, and she would walk the Green Mile. There was no way he was going to buy the 'TIA' defence. Still, she had to try *something*, but what? Maybe a front-foot, faith-based approach: *I've been trying so hard to let God into my heart, baby. Remember, me converting was your idea. You told me to trust Him no matter what. The other night, I truly felt Him compel me to commune with the two boys across the hall the way we did [twice]. And, although I found it odd that He specifically asked me to scissor with the German girl in room 207 last week, I obeyed. You'll have to take your concerns up with Him, not me.*

But their exchange never got to this point. In an instant, several lifetimes' worth of luck were cashed in on her behalf. Before she could begin, he interjected ... to give a confession of his own. He could no longer live with himself, he said, if he continued to lie by omission. He had to speak His Truth: a few weeks ago, after having one too many drinks at a party with his buddies, he had 'accidentally decided' to fellate one of the frat brothers.

The timeline leading up to the blowjob was described as a 'grey area' and, tellingly, he gave no information *whatsoever* as to whether the act was performed to completion. My hunch (which I kept to myself) was that our theologian had gulped down a 'vaccine', Sambian-style, of copious volume. Incredibly, the couple ended up getting married and, within a year of her return, had a kid on the way. God gives the toughest battles to his strongest soldiers, I guess.

THE PEARLY GATES

After our final exams, the French drove north into the Namibian desert region called the Sossusvlei, home to famously large sand dunes (some of which were over three hundred metres tall!). I was invited, but it would have required being trapped with six of them in an all-terrain truck for twelve hours straight. *Twice.* I opted instead to join a few of the American guys along the Garden Route. They wanted to hit the usual tourist stops, but had also been invited to an end-of-year party in Mossel Bay; one of them had a family friend with a small holiday home there, and my room-mate Sean and I had weaselled our way onto the guest list. With the weather was getting warmer, we thought it would be nice to be near the water. Not that we were intending to swim; the bay was grimy and industrial, not good for much beyond shark diving (for which I'd explicably agreed to join them the next morning).

Sean and I killed time on our drive by trying to plan for any eventualities we might face over the course of the evening. We didn't know the other guys too well; they lived in a different dorm block and took separate classes, but we were excited nonetheless, as we'd been told that the town had decent night-life and was a hub for other young travellers exploring the coast. In our optimism, the only problem we envisaged facing was how we would juggle all the different options: sorority houses, pool parties, strip-poker tournaments, and so on. Being so in demand was as much a blessing as a curse, you see. 'Ugh, you're going to *hate* us,' we'd have to say to the Swedish bikini team, right as they finished inflating the blow-up pool for naked mud-wrestling night. 'But we've gotta run. It's just that we signed up to be judges for a stupid wet T-shirt competition down at the beach, and it starts in a few minutes.'

We arrived after sunset to find that reports of a 'beach house' had been grossly exaggerated. It was only a single-story, two-bedroomed caravan, and it was already packed full of bored, musky young men. Somebody had brought three extra friends from another school; there was no air-conditioning, and, in their humidity-induced delirium, the sweaty group were now lounging around in their boxers. As if they'd been waiting to be spurred on, they sprang into action upon our arrival. We divided into smaller groups and scattered toward the shore.

The evening, until now merely anticlimactic, was complicated an hour or so later by a voicemail from our host, who informed us that a heavily intoxicated member of the shirtless division had procured a willing female, brought her back to the house, and proceeded to make sweet love to her on the back porch. Their coitus was so protracted, and the participants so vociferous, that two of the neighbouring houses had been awoken, with their elderly occupants threatening to call the police. Perhaps due to the stress, the girl had then projectile-vomited repeatedly over most of the porch, including the barbeque grill and some of the house's back wall. Her gallant lover then offered to help her clean up the mess. With no clothes of his own at hand, he ran inside, grabbed the first cloth he saw, and mopped up as much as he could, before falling asleep next to her on the little patio. It turned out that our modern-day Sir Walter Raleigh had used a favourite ornamental blanket of the host's grandmother to soak up the vomit.

The voicemail ended with us being told, in no uncertain terms, that we needed to find somewhere else to stay. We were happy to oblige – we were so horrified by the story that we'd have been far too ashamed to hang around – but this left us in quite the dilemma. My first thought as we came back to the house to pack up our things was to build on the foundation laid by our fallen comrade: 'Wow. The puke girl sounds like a real piece of work. And I'll bet she had a bunch of friends, perhaps just as crazy as her ... Just out of interest, where on the promenade did he find her, exactly? That way we know where to avoid.'

This plan didn't work. Not because it wasn't ingenious, but because the would-be informant was non-verbal from intoxication. So Sean and I set out on our own reconnaissance. I suppose we *could* have paid for a hostel room somewhere, but it was now nearly midnight; we were in too deep. We made our way along the promenade for an hour or so, but were unable to make much headway. A change of tack was in order. We went for the 'lost dog' strategy: rather than waste energy covering territory, stay put and let them come to us.

Spotting a free bench, we sat down to put ourselves in the shop window. Depending on the look of the groups who eyed the bench as they passed by, we indicated we were saving the other half for soon-arriving friends – a well-worn tactic by any regular patron of public transport. Before long, a group of American girls approached. There were four in total, two of whom were quite pretty and seemed to be leading the others.

'Can we sit here for a second?' one of them asked.

'Absolutely!' we said, making space for them. But she'd clearly been referring to the two bigger units, who had just caught up and clearly needed to take a load off.

'We'll catch you guys later,' the first girl said to her now-seated associates. 'It was nice to meet you!'

The two pretty girls hurried away down the boardwalk. Sean and I grimaced at each other; the hot potato had been expertly passed. But, as we overheard one of them complaining to the other, our ears perked up.

'Ugh, that guy from earlier keeps texting me.'

'Who?' the other one asked.

'Rory. Maybe we *should* go over. He said they made a ton of jungle juice.'

This was our window. Pretending Sean had made a joke, I laughed loudly, hoping to get the girls' attention, and also to show that we were fun-loving, trustworthy guys. It seemed to work.

'How's your night going?' asked the girl next to me, turning to face us.

'Better now!' I said. 'But if we sit here any longer, I'm going to fall asleep. Let's go for a walk.'

As we wandered down the street, we told them we'd be happy to escort them to the party.

'You really shouldn't trust random dudes,' we added. 'It's a crazy world out there. We can't just let you wander into a lions' den.'

They happily agreed. *Perfect*: we'd complete this little side-quest, then crash on the girls' hotel room floor, and sneak out before they woke in the morning. Talk about failing upward; this was what the American Dream was all about, baby. Fortune favours the brave (and the scummy).

We arrived at the party to find there were only two other girls there, hilariously outnumbered by almost two dozen guys. Sean and I mingled for a little before conferring, deciding to ask the girls if they were ready to leave. I caught the eye of one of them and walked toward her from across the room. She was about six feet tall, black, and named Heaven. Naturally, she had a large tattoo of a skull on the front of her upper thigh. Before I could speak, she grabbed me by the shoulders and thrust me against the wall.

'Let's quit fucking around. I know we've been trying to pretend there isn't something between us. Here's what's going to happen next: you and I are going to go back to my room and you're going to fuck me.'

'That sounds awesome, Heaven, and I'd love to. But I'd hate to ditch Sean. Why don't I just check on him to see if he'll manage on his own?'

'Oh, don't you worry about him,' she answered. 'Lauren's had her eyes on him all evening.' I looked over her shoulder and across the room, where I could see Lauren. She did seem to have *someone* pinned in a corner, but her frame was so large I couldn't tell who it was – I'd have to take Heaven's word for it. Mercifully, we were interrupted by a commotion in the centre of the room. The party's host, Rory, diminutive but confident, was deep in impassioned conversation with a much taller, leaner, and heavily tattooed young man.

'I love you so much, bro,' Rory said. 'But listen. Even though we know each other, I still feel like we need to actually *bond*. We've gotta fight. It's the only thing that will connect us. We have to do it now – I leave for the army next week.'

'Bro, I'm telling you, it's a stupid idea,' said the friend. 'You *know* I'm training to be an MMA [mixed martial arts] fighter.'

'I told you, I don't care if I lose, it's not about that. What if I never see you again?'

'Whatever, dude. If it's what you really want.'

Sean (who'd manage to wriggled away from Lauren) and I had no idea what to do. Even if we'd wanted to, we didn't know anyone else well enough to intervene, and we couldn't risk getting kicked out now that we'd been so explicitly propositioned by the girls. Whatever the circumstances here, they seemed far preferable to what might happen in (and under) Heaven. Several of their other friends tried to separate Rory and the fighter, but this was futile. The two were sober, or near enough; there was no animosity, no heightened emotions to diffuse. Unconsciously, we formed a circle, as if preparing for a dance-off. Given the course of the night, it should have come as no surprise to us how quickly the two combatants disrobed to their underwear. The MMA fighter had a single-digit body fat percentage and CHEROKEE WARRIOR tattooed across his back. They did some stretches before a volunteer referee stepped in.

'Just regular boxing rules, right?'

'Yeah, obviously,' said the WARRIOR.

'Fuck that,' said Rory, 'I want everything. Grapples, take-downs, all of it. Anything goes.'

They shook hands, and then started to spar, each landing a few glancing blows. Rory was clearly burning energy trying to stay agile and light on his feet; I hoped he would tire out and surrender. It was obvious the MMA fighter was keeping him at arm's length, but Rory kept throwing harder punches, and soon started to kick out. Presently, he landed one of these on the fighter's calf. The larger man winced in pain and decided their fun was over. He spun, producing a brutal roundhouse kick that slammed into the side of Rory's head and instantly knocked him unconscious. The victor didn't celebrate. You could tell he wasn't necessarily proud of his work; all he'd done was subdue a pesky subordinate. It was reminiscent, I realized, of the Hawaiian on my last trip. *Had I missed some sort of memo about male bonding?* If 'aggravated assault' was what the bar for friendship was nowadays, no wonder so many young men were feeling lonely.

Rory, who'd crumpled to the floor from the impact, woke up after a minute or two. He went over to his friend, gave him a hug, and thanked him for the closure he'd been so desperately seeking. While relieved for their friendship's sake, we were more grateful the house hadn't become a crime scene. With the host's survival at least temporarily assured, we could hang around and pretend to fall asleep on a couple of the couches to avoid having to leave with the girls.

෪

Between the sleeping arrangements and the looming prospect of next morning's sharks, I didn't get much sleep. I agreed to go through with the dive but decided to leave my camera back on land. This proved wise. This time, a different shark (her head much wider than the previous one) recreated that terrifying approach I'd watched hundreds of times, surging out of the depths and slamming into our cage. Without the distraction of my camera, I could see the attack coming and dodged it easily. Unfortunately for her, however, she not only missed the tuna but bit into the chicken wire at such an angle that she couldn't dislodge herself. For two interminable minutes, we were trapped inside the cage as she desperately tried to thrash herself free. At long last, the crew managed to lift the roof and help us scramble out. We watched on as the captain sat on the edge of the boat and tried to pry her free by pressing his foot against her nose. It was a heartbreaking sight: she hadn't meant to crash into us and was clearly panicking as she slowly drowned. Finally, the captain succeeded. She slipped away, exhausted (and a few teeth lighter) but alive. I had no desire to push my luck further. As soon as we were back on land, I drove home to the university, leaving Sean and the other guys to continue their road trip without me.

But my partying days weren't over just yet. A few weeks later, the American girls talked me into joining them for a music festival a couple of hours north up the coast. They offered to split the cost of the rental car amongst themselves, as long as I agreed to drive (none of them were comfortable enough with stick-shift). This kind of thing wasn't usually my scene, but I vaguely knew a couple of the performing artists and the tickets were cheap, so I agreed on the condition that we didn't leave until Friday – I refused to go for all three nights.

TWELVE HOURS
A RAVE SLAVE

The drive north was stunning – although the first leg was actually *east*ward. The girls had wanted to go on a safari while in the country, but weren't sure they'd get a chance to do a trip along Garden Route, so they made reservations at the only reserve within 100 miles of the cape. It was more than an hour out of our way, but we had a rental car and an afternoon to kill.

While the view from Sir Lowry's Pass, on my first visit to the Cape, seemed to celebrate what was behind before looking beyond, the Huguenot Tunnel, which passes through the Du Toitskloof Mountains, does the opposite: *Forget what you thought you knew.* There is an old pass available, in the mould of Sir Lowry's, but the new tunnel is quicker and safer, especially on a day like this one, when the roads are wet. The jagged, alpine panorama would be impressive enough without the dramatic reveal, of course, but the several minutes of sensory deprivation make the exposition that much more breathtaking. After a few miles, these more conventionally shaped mountains recede, giving way to massive, angular chunks of rock that erupt through the earth.

The mountains back in Stellenbosch were noteworthy for their horizontal stratification, their distinct layers illustrated their steadfast passage through the eons. By contrast, the landscape here seemed to celebrate – even insist upon – its noncompliance. The passage of time, it declares, need not be synonymous with gradual erosion and decay; there can also be spontaneous creation, however imperfect, however chaotic. Indeed, the fold aspect of the Cape Fold Belt, as the region's mountains are officially named, becomes easier to conceptualize; like the mountains nearer the Cape, these formations, too, are layered, but at any angle to the earth *except* horizontal. So, on the road goes, navigating through a graveyard of abortive fragments of accordioned crust.

The sense of being in a cauldron of tectonic experimentation was heightened by the weather. It was mid-morning, but the sky was a leaden grey, with billowing cumuli casting the landscape into foreboding shadow, and the air carrying an unmistakable pre-storm charge. The intermittent lightning strikes we'd seen in the distance were now surrounding us, and were accompanied by sideways rain. After a few minutes, the game reserve called to tell us the downpour had damaged one of the elephant enclosures, and that the safari drive had been postponed to later that afternoon. The girls decided they'd rather call it off and head to the festival so that we didn't have to drive at night. We left the inclement weather behind us as we made our way northwest through the farmlands.

᠔

We were one of the only groups who arrived late. The parking lot, a mown field adjacent to the festival grounds, was already filled with thousands of cars, so we had to hump our gear almost a mile to the entrance gates. Even the short queue there ended up taking about an hour as a security team thoroughly rifled through every last backpack, duffel, and sleeping bag. This was, ostensibly, a safety precaution, because they didn't want anyone bringing in glass bottles that might be thrown or broken. Coincidentally, there was supply to meet this newly created demand: 'safe' alcoholic beverages served in plastic cups and aluminium cans were provided at exorbitant prices by the festival's sponsors.

After we'd been all but cavity searched, we were waved through and given a map of the grounds. There was a main central stage and a handful of smaller ones scattered around it; there would, the brochure proudly informed us, be performances on one stage or another straight through to Sunday afternoon, allowing us to 'rock 'round the clock.' Interspersed in the music zone were clusters of food and drink trucks, and other social and lounge areas. During the day, these would host group activities, such as yoga, art, and carnival games.

One of the workers looked at our tickets and circled a spot on the map where we could go to pitch our tents. There was, apparently, a bit of space left in Zone G. We made our way around the edge of the grounds to the general admission campsite. Sleeping outdoors isn't something I pine for even when conditions are ideal, but here we'd be surrounded on all

sides by other tents, with only a metre-wide access path leading to the stage area a hundred metres (109 yards) or so away.
There had been glamping options available; larger, heated
tents with en-suite bathrooms and warm showers, but these
had sold out long ago. All that had been left to reserve were
patches of grass, and it was up to us to bring tents, sleeping
bags, and other supplies. The nearest town was a half hour's
drive away, so retreating to a hotel or bed-and-breakfast was
out of the question.

If we'd thought we'd be arriving before the party really
got going, we were mistaken. The campsite looked, if I
had to approximate, like a post-natural disaster emergency
camp. This was a joyless, hopeless place. All we were missing
were a few pairs of blue-helmeted UN peacekeeping troops
patrolling the perimeter. Flags on ten-metre poles were dotted
throughout, marking the dozen or so different zones. We
shuffled, fresh-faced and wide-eyed, toward our designated
spot, directed by brusque workers in neon shirts stationed
throughout the campsite to guide the lost and weary. It felt
as if we were replacing shell-shocked, trench-footed soldiers
who'd been holding the line for weeks.

We passed a large first-aid tent, one of several, according
to the map. We couldn't see inside, but streams of fraz-
zled, harried-looking attendants and medics streamed in
and out. A line of non-emergency casualties waited outside,
occupying states between doubled-over / dry heaving and
face-down / unresponsive.

'Move. Move! Coming through!' We were shouldered out of the way by a stocky woman making way for two of her colleagues, who were carrying a catatonic girl on a stretcher. The patient wore nothing but a bikini and hiking boots. Instinctively, we stopped and stared, hoping for a sign of life. We got one: she turned her head slightly and mumbled something. Before anyone could respond, she belched out a few pints of vomit, soaking her pigtails. One of the Zone G workers noticed us.

'Alcohol poisoning. Mild. She'll be fine.'

We traipsed onward. The faces we passed on the way to our tent showed varying degrees of haunted, jaded, sallow, and gaunt. This made me feel all the more ridiculous: as we'd waited in line outside, I'd let the girls adorn my cheeks and forehead with glitter, which they'd mixed with Vaseline to make sure it would stick. The concoction happened to be waterproof, not that this would come into play – rain wasn't in the forecast, and there was no hope of finding a shower. For those not glamping, the long weekend's worth of accumulated grime and musk was to be regarded as almost a badge of honour, or so I was told by the girls, who did this sort of thing (both the festival and the glitter) regularly.

I found the romanticization of 'slumming it' ironic. A vast majority of the festival attendees were white – tickets for four-day access were a couple of hundred dollars, food and drink not included. We'd driven out into the remote farmland to set up our own little township and live, for a couple of days, in squalor, just like the people whose settlements we toured/gawked at back on the Cape. The appeal of such an event clearly went beyond the music: this three-night Saturnalia was a chance for us to roll around in the dirt and disavow hygiene with no social consequences, secure in the knowledge that we had the safety net of paid staff on hand to rescue us if things got out of hand. Here, grunge and recklessness were purely performative, a spectacle in their own right. They were subversive, an act of rebellion. We were not regressive and barbaric, but liberated and authentic.

At any rate, I'd have to shelve my qualms for now. We were here, so I knew I had no choice but to go along with it all. *I could last a couple of days, right? A bit of dirt never hurt anyone.*

But a bit of dirt this was not. Foolishly, I'd taken the glitzy marketing pictures of the festival at face value. I hadn't accounted for what sufficiently heavy foot traffic will do to a perfectly healthy, dry grass field. The mass campsite was so cramped that the thin paths dividing our tents had, in less than a day, been churned up into squelching, muddy ruts. It seemed that everyone else, including the American girls, and even the overzealous (and possibly dying) bikini girl, had prepared for this, bringing heavy-duty hiking boots – except for me.

We found our allotted space and set up our tents. Miles from the stage area, but pretty close to the line of two-dozen portable toilets near the outer boundary of the festival. *Win some, lose some.* Two of the girls had gone halves on a fancy, queen-size air mattress and spacious tent. Given that I was a last-minute addition, I was to share a tent and blow-up mattress with the remaining girl, who happened to be the one who liked getting back at her boyfriend back home. I know what you're thinking, but I knew a minefield when I saw one, and I remembered what Chet had said about where *not* to stick it. (Although that seemed to be a case of 'Do as I say, not as I do.') Beyond that, the sheer amount of filth and the total lack of privacy made the ambience a distinctly unerotic one.

But apparently I was in the minority on this point. The close quarters, drugs and alcohol, combined with hysteria born of sleep deprivation seemed to create some sort of aphrodisiac. Sure enough, my bed-mate fell victim to whatever pheromones were floating around, making out with a guy she met later that evening before leading him back to her (our) tent to consummate proceedings. For whatever reason, she changed her mind at the last second, opting to give her nameless lover a hasty rub-'n'-tug behind the toilets. *What about the layer of dirt?* – you ask, earnestly – *Surely, there would have been too much friction.* You're right. Luckily, our girl was a quick thinker. I found the smoking gun on the blow-up mattress when I returned later that night: the tub of glitter-infused Vaseline, from which a fistful had been hastily clawed.

And why had my night been reduced to picking up after glitter girl? Well, it was because I'd tried to solve racism. No good deed goes unpunished.

The evening had started off fine. Recognizing the name of a band (Two Door Cinema Club, an English indie group who were big in the early 2010s), I made my way over to one of the smaller stages where they'd be performing. A few minutes before they were scheduled to come on, I consulted my trusty map to see if there was a place to grab a beer. There was, just a few minutes' walk away. I wandered over, but saw there were already a few dozen glitter-coated rave-slaves waiting in line. On top of this, one pint of generic lager cost fifteen dollars. *Forget it.* Luckily, we'd chugged a few 'cocktails' back at the tent in preparation (old faithful: half knock-off-brand energy drink, half industrial-strength paint thinner masquerading as vodka), and these were now taking hold. Although I now had to make a concerted effort to walk in a straight line, I made my way back toward the stage. I called one of the girls along the way – no answer.

I was then stopped by a jovial-looking guy walking in the opposite direction, who offered me a high five. If it had been anyone else, I'd have kept walking, but the pressure was on: he was black. This was my moment to give back, to display to him (and any witnesses) that I was a proud ally. I returned the high five, hamming it up and putting my whole body into it. This did the trick: he roared with laughter. *See? This 'activism' stuff wasn't so hard.* I'd made this guy's day – likely his whole month – and what had it cost me? *Nothing.* He came back for more, offering another high five, along with one for the other hand. I obliged. He then mimed that I was to follow his lead – a game of interracial Simon Says. We did the two high fives simultaneously, then I followed him through a little dance (à la the Macarena), ending with hands on gyrating waist. It was just like that scene from so many movies: the charismatic black guy (with his innate sense of rhythm) showing the hopelessly wooden white guy how to dance. *It's all in the hips, my man. There ya go, just like that.* We celebrated with a hug. It gave me immense pride to have built this lifetime bond. At long last, I could say I'd earned the respect of a *real* South African – a box I'd not yet been able to check, and certainly wouldn't have been able to back in Stellenbosch. I thanked my new 'homie', and wished him well as he sauntered off into the night. I hoped that he took something from our encounter, too.

He sure had. I pulled out my phone to try calling my friends, desperate to find them and brag about my achievements. Rather, I *went* to pull out my phone, whereupon I realized it wasn't in its usual pocket. Nor in any of the others. I scanned the grass around my feet. *Nothing.* My stomach plummeted as I realized what had happened. *And the one time I'd dropped my guard, too.* Thankfully, after a few minutes I managed to spot the girls, and waded through the crowd to fill them in (I cut straight to the part where I was victimized by a good-for-nothing lowlife). The concert had been ruined, as had my mood, so I wandered back to the tent.

Literal rock bottom came a few hours later when I awoke on the frozen ground at four o'clock in the morning. Evidently the air mattress had been microscopically punctured at some point. Happily for them, a couple in the tent next door had suffered no such misfortune and were making the most of their evening. It was difficult to hear over the thumping bass from the nearby rave tent, but it sounded (based on her remarkably loud slurping) like she'd smuggled in some sort of ice pop. By the sounds of things, her companion was taking immense – one might even describe it as *orgasmic* – pleasure as he egged her on. *A bit cold outside for frozen desserts,* I thought, *but to each their own.* I managed to bundle up just enough to fall back asleep, but *not* enough to prevent the excruciating back pain I awoke to, and which plagued me for the next two weeks.

For what it's worth, the pain wasn't what actually woke me up. That honour went to the heat. The cheap membrane of the tent had acted as a sort of magnifying glass for the sun, with me as the insect or slug to be burned. I woke with an unholy gasp, as if I'd been dragged unconscious from a burning building and brought back with smelling salts. After a minute or two of thrashing around, I managed to extricate myself from my sweat-sodden blankets and fight my way into a seated position. From there, I could yank down the tent zipper and let in some fresh air. It was only *fresh* in that it was of a lower temperature, not by any other metric: at some point in the 'wee' hours of the morning, someone had decided they weren't going to reach the toilets and had settled for using my tent's doorstep to empty their bladder – along, for good measure, with the contents of their stomach. These two puddles could have been left by separate offenders, but I believe someone willing to do *either* would have been willing to do both.

That afternoon, the festival posted on their official social media accounts that they'd received multiple reports that a gang of pickpockets from a township outside the closest town had managed to spoof tickets and break into the grounds with the sole intent of taking advantage of the relaxed, communal atmosphere and, more specifically, the lubricated, vulnerable masses. *Just in time, fellas, thanks for the heads up.* Luckily, the theft of my phone was to be a victimless crime: I had insurance, and a friend of my parents was visiting Cape Town at the end of the month for work and could bring me the replacement phone. I'd helped the underprivileged urban community and could palm off the cost to the money-grubbing insurance industry. Win-win.

When my new phone arrived, and I slapped in a SIM card from my South African carrier. It turned on, but could not be activated, saying that I was out of the country. Supposedly, iPhones were no longer region locked – a selling point had been their international compatibility. I called my carrier back in the US. They not only denied that the phone had been region locked but that it even *could be* locked. They humoured me regardless, sending a special signal to the phone via its unique device ID.

'There you go,' the man said, 'we reached your device and can confirm that it's been activated. Will that be all the help you need for today?'

The device may have been telling *them* it was activated, but that wasn't the story it was giving me ('This phone cannot be activated.'). I then called Apple, who also vehemently denied that the phone could be locked. Apparently, I'd been sent the first phone in history that was not only sentient but mutinous. I began to wonder if this was one of those depressing stories you hear about where a sweatshop worker tries to cry out for help from the outside world. If that was the case, they sure had made their message cryptic enough. Or was the medium the message? As in: *though the System may seem all-powerful, it's more dependent on the individual than it would have us believe.* Perhaps. But now wasn't the time – I needed a working phone. Apple suggested I visit one of their stores, where they'd apparently be able to connect my phone to some magic computer that could activate my stubborn device by brute force. *Finally, some actionable advice.* I looked up the nearest option, assuming they'd have a store in Cape Town. They did. I didn't take note that its name was not Apple but something adjacent: iStore or iApple or the like.

What this meant, I learned, was that it was not an official Apple Store, but a licensed vendor. South Africa, it turned out, had no actual Apple Stores. Not because Apple wouldn't have liked to set up shop, but because of a government policy called Black Economic Empowerment (BEE) that requires any company hoping to peddle their wares to prove that they meet a litany of different requirements and quotas – mostly regarding the racial make-up of its employee and management groups. BEE, along with other positive discrimination policies implemented to redress the imbalances of apartheid, are what allowed certain ANC chiefs and black businessmen to become obscenely wealthy. When nationalized industries were divvied up to the private sector, a bidding group with a majority-black ownership was favoured over one with white leadership. Unsurprisingly, this was – and still is – exploited by companies that may fulfil these quotas on paper, but have little interest in contributing to societal change.

Not that I gave a hoot about any of this at the time. I was indignant and couldn't fathom that, in this day and age, a phone could just decide to lock itself and have that be the end of the matter. In the end, I had to cave in and buy an old, used phone from the sort-of-real-but-not-actually-real Apple Store.

I'd rented a car for a few days to do all this running around and still had a day left, so I decided to use it for a revenge mission.

PIPE BOMB

The springtime music festival we had attended was called Rocking the Daisies, so named for the few weeks each year when the country's wildflowers come into bloom, producing a magnificent spectacle, with acres of polychromatic carpet covering the rolling grasslands and hillsides along the coast. Supposedly, that is. They certainly weren't in bloom while we were there. The landscape was still pretty, but nothing you couldn't have found back in Lancaster County. And so I resolved to track these elusive flowers down. Their mecca, according to a couple of my professors, was Postberg (pronounced *Pot*-berg) Nature Reserve, an hour up the western coast. As I waited in the queue of other cars outside the reserve, my excitement grew. It was surely a good sign that the attraction was in such demand.

Alas, as I entered the park and swept down into each meadow, I saw that the gentle hills of grass overlooking the ocean were still just a regular, healthy green. As I crested each hill, I expected the treasure trove to finally reveal itself, a dense display of flowers transforming the hills into a gleaming horde of gold in the sunlight. But a smattering of daisies was all I got – and even these were just plain-old white, which doesn't even count.

The terrain itself was interesting, at least. Like so much of the peninsula, it was littered with boulders and rocks. This made sense around the Cape given its steeper hills, which debris has tumbled down over the millennia. Many of the local beaches were scattered with boulders on the shores, giving shade to sunbathers, or, on one secluded little beach, to *penguins*, where a colony claimed a particularly rocky inlet along the peninsula called Boulders Beach. The knee-height birds that waddle around and burrow into the nearby wooded undergrowth showed up in the 1980s – and they might not be there for much longer, having been classified as critically endangered after years of industrial over-fishing in the surrounding waters.

Anyway, on the flatter surfaces, these rocks make for a more curious, incongruous sight, especially at altitude. Hundreds of feet above sea level, stretches of road navigate the geologic handiwork of ancient glaciers, winding through acres of low grasses interspersed with chunks of light grey rock that range in diameter from several centimetres to several metres. The smaller rocks are often spiked, unrefined, volcanic-looking, while the larger ones are smoother, oblong, ovoid, or baguette-shaped – though rarely perfectly spherical. All sizes are seen stacked on top of each other, not necessarily in any order, balancing with impossible poise against the unrelenting elements.

After snaking through this almost lunar landscape for about an hour, I passed a small parking lot with a gazebo and some wooden picnic tables next to a beach access path. This seemed like the perfect spot to rest for a few minutes before I began my drive home. I sat and started to unpack the lunch I'd brought, enjoying the breeze coming in off the Atlantic. It was a cool, blustery spring day, but I'd come prepared, having grabbed a thermos for a few dollars from a grocery store just before setting off. I was proud of myself, having brought a couple of teabags, filled the thermos from the kettle, and even remembered to fill the little extra compartment with milk. I was heartbroken, then, when what emerged from the thermos was a molten slurry of shattered glass. In bouncing over the speed bumps placed along the park's road, the impact must have caused the vacuum to implode. There was no salvaging the situation, as the shards were so tiny they'd shred my oesophagus if I took a sip.

I noticed a few budding flowers running to my left and continuing out of sight behind a pair of boulders. After carefully resealing my newfangled pipe bomb, I ventured over to where I prayed the park's only colourful flowers had hidden themselves. At this point, even just a few dozen would be enough to take something from the day. Sure enough, a small cache of orange and red flowers had sprouted behind the rocks. A triumph, all things considered.

Wanting to stretch my legs a little before getting back in the car, I decided to walk along the shoreline that stretched for miles southward toward Cape Town. I left my socks and shoes by the flowers, but I soon regretted this decision. The sea foam was a sickly, synthetic green, and the seaweed heaps I'd noticed from back at the rocks were actually composed mostly of gangrenous flotsam. The breeze had turned slightly rancid as it filtered over and through the mounds of debris and rotting kelp. It was time to get out of here. I picked one of the dark lumps up ahead at random, resolving to turn around after reaching it.

As I approached, I saw that it was actually a young seal lying on its side, lifeless. The sand caked deep into his tufts of moulting fur indicated he'd been at the mercy of the elements for a good while: the tide had rolled him toward the grassy dunes before retreating to let the sun bake him dry. However peaceful his current state, the thick tangle of heavy-gauge fishing line around his neck made clear that his end had been wretched – the line had nearly decapitated him. There seemed no other evidence of grave injury, all that surrounded him were hundreds of bird footprints. Rubberneckers, I figured, before noticing that his eyes had been pecked out. I hoped this feeding frenzy had happened *after* he'd died.

I looked out to sea, where a handful of fishing boats trawled along the western horizon, then made my way back to the picnic table. I tried to have a few bites of my sandwich, but I'd lost my appetite. It was impossible to know for sure, but there was a non-zero chance the seal had been killed by the same net, trawler, or fishing conglomerate as the tuna in my sandwich. Did this somehow make the seal's death more of a tragedy than the tuna's? If so, why? Mammalian solidarity?

In the abstract, this process of relentless globalization is glacial and, unless one spends time at its bleeding edge, easy to forget. It's much nicer when we don't have to see how the sausage is made. Our impossibly complex, international supply chains completely alienate us from the products we consume, even fundamentals such as the food we eat and the water we drink. We don't have to know or care where anything really comes from, let alone what it costs. But the thread of culpability exists nonetheless. As information about all this becomes more available, the 'ignorance' defence grows more and more dubious.

The real kicker is that it doesn't matter either way, because we also know we can't participate, or even exist in the modern world *without* consuming. And so we are forced to be hypocrites, supporting a system (if not in word, then in deed) that we know to be not just inhumane but utterly untenable, demanding infinite extraction from a finite planet.

The mutilated seal was by no means the first innocent victim in these parts, nor, I fear, was his fate the most unsightly. For as untamed and unassailable as the region's natural landscape appears, it has not been spared the slow creep of human expansion. Before we brought in the rifles and bulldozers, the Cape and its mountains were home to many animals, including lions, hippos, and wild buffalo. A handful of leopards still inhabit the region farther inland. Today, all that's left on land are the baboons.

Well, that's not entirely true. There's also a population of little creatures called *dassies* which, despite their resemblance to groundhogs, are, somehow, most closely related to the elephant. They received none of the majesty of their distant cousins, however, instead appearing troubled and slightly frumpy. Their dark, beady eyes are framed by a tufted, furrowed brow, while a small, mammalian snout hangs over a thin-lipped, jowly mouth. They make for harmless hiking companions, as they're furry and nonviolent, and their semicircular ears are somewhat endearing – cute would be too strong a word.

Anyway, as I climbed Table Mountain a month or so before I left the country, I was joined by another of the region's creatures supposedly consigned to oblivion, a parable of human tampering in their own right. For the first section of the trail, I trudged up a gruelling incline of rocky steps directly under the track of the cable car, every few minutes smugly giving the middle finger to the gondolas floating overhead on their effortless beeline to the summit. After a thigh-burning hour or so, I split from that trajectory and was out of their view. The slope became almost a sheer face and, after a few minutes of hand-over-hand climbing, I rested on a rock ledge looking out over the City Bowl. And that's when I saw him: to my right, on a small outcrop serviced by no visible trail, a lone mountain goat, stock still, contemplating the lattice of streets below. The thing is, mountain goats weren't endemic to South Africa.

Later that evening, I did some research and learned these goats had a curious backstory. A zoo had been set up by Cecil Rhodes just outside the city on the western slopes of the Cape mountains, right next to the University of Cape Town (UCT) campus. It had housed lions and other animals, including a handful of specially imported Himalayan goats. Sometime in the early twentieth century, a pair of these had orchestrated an escape into the forest behind the zoo and had decided to get straight to work, romantically speaking. Given the health issues the Amish struggled with back home, I shuddered to think of the birth defects incurred here, given the repopulation committee was made up solely of our cloven-hooved Adam and Eve. This wasn't so much a gene *pool* as a gene *shot glass*. Luckily for the prolific couple, their genome proved resilient – so much so, in fact, that the local government decided to exterminate the ballooning population. A few stragglers had managed to evade the hunters and still lived on the mountain, but sightings were rare.

But questions remained over the goats' original home. I'd never heard of the zoo before, and exact details proved tricky to find. Nobody at the university knew exactly when or why it had been shut down, but most rough estimates were that it had been abandoned in the 1970s or 1980s. There were suggestions that several UCT students had broken into the zoo at night and had managed to disturb, and subsequently get bitten by a lion, which may have caused the suspension of operations – some even claimed it had been students who'd let the goats out those decades before.

In any case, the zoo's ruins had been left open for explorers, and I felt compelled to visit out of respect for my hiking buddy. From the campus, I followed a quiet trail beyond a car park and through a wooded area, eventually coming upon the outer edge of the zoo. Like any abandoned property, it was eerie, even on this sunny afternoon. Cement terraces, where there must have been pens or cages for the smaller animals, were levelled into the gentle slopes at the base of the mountain. Above these, I came upon the largest exhibit, which had housed the lions. A tall chain-link fence surrounded a semi-circular moat, a few metres deep and perhaps two across. The inner peninsula had several platforms and small trees for the lions to climb or rest under while the crowd outside the fence watched on. I ventured around behind this island to the lions' private enclosure, where I realized I could actually squeeze in. The individual cages each had a thick metal guillotine door that slid up at feeding time; rust now kept these stuck open or had eaten them away entirely. Stooping through these claustrophobic passageways, whose grim trapdoors and rusted grates let in only brief flickers of light, it felt as if I were beneath the Colosseum, waiting to meet my fate – all that was missing were the baying, bloodthirsty masses. Finally, the tunnel reached a wrought-iron portcullis that opened out to the island. I'd have preferred to climb the fence rather than go back the way I came, but the moat was too wide to jump.

The concept was depressing: a zoo had been erected to display a lion shipped in from abroad because the local ones had been slaughtered. Realizing the game drive I'd enjoyed was based on a similar premise, I took comfort in the selling-point that had previously been so unconvincing: the private reserves I'd visited had sheltered and rehabilitated traumatized animals. So, I was nothing like those zoo-goers of yester-year, I reassured myself. All told, I'd paid for weeks, *months*, of elephant, and lion, and springbok rehab – not to mention the help I'd given to the country's sharks. I'd done my bit; my conscience was clear.

RED TAPE II

I spent the next few weeks studying for my final exams. It felt like they went pretty well, but when I received my final grades, I was horrified. I spoke to a few of my teachers and was relieved to learn that in South Africa, grading was done on a bell curve: an A or A+ (90 or 95 percent) was almost impossible and B's or C's were perfectly commendable results. My home school, who'd already been looking for an excuse to give me a hard time, didn't believe a word of it. Eventually, I convinced the registrar in Stellenbosch to draw up an equivalency chart. This seemed to placate the Americans, who sat with it for a few weeks before admitting defeat. This delay meant my graduation was an anticlimactic one: in mid-February, about two months after returning to the US, I got a terse email from the administrators conceding that I had [to their chagrin] graduated. (Attached was an alumni fundraising form; the school was hoping to build a new gymnasium.) But never mind the minutiae, celebrations were in order: I was now a college graduate! *Huzzah!*

This was, in fact, a serious problem. The thing was, I'd never thought this far ahead. My undergraduate years had been spent thinking only about my next assignment or test – and how I could pass with as little effort as possible. But I'd now run out of road. The only thing in front of me, it seemed, was the corporate hamster wheel. I had no interest, but the alternatives weren't too appealing. Farm country didn't have much to offer beyond agriculture or working at one of the Amish puppy mills outside the city.

For a few months, I worked as a busser clearing tables at a local restaurant. It was only minimum wage, but I was paid more than the actual waiters (in the United States, restaurants only need to pay them about two dollars an hour, the rationale being that they'll end up making more than enough from tips). I was entitled to one cent from every dollar they were tipped, which could add up to being ten or fifteen extra dollars per shift, but the catch was that this had to be collected manually, so I had to track each of them down before their shifts ended and beg them to give me my cut. Regardless of how much they had made that evening (sometimes well over a hundred dollars), they were unfailingly tight-fisted: it was common to be given a dollar bill by someone who owed me, say, eighty-seven cents, and told to find the change.

When I complained about this at Thanksgiving, my grandpa, who'd worked as a state game warden and had been retired (with a state pension) since the Reagan administration, told me I should just do what he did as a kid and pick up a summer job trapping beaver up in the mountains. I politely explained that this might not be viable (and hadn't been for nearly a century). My dad then threw in his two cents, suggesting that I become a door-to-door egg salesman: 'Here's what ya gotta do. Go to a couple Amish farms and buy up a few cartons at a dollar a pop. Then go around town and flip 'em for two or three. Easy money. That's what I did all through college – graduated without a penny of debt.'

Thankfully, an old tennis coach reached out, sparing me from the world of egg arbitrage. He was taking a sabbatical and asked if I could take over his client list for a couple of months. I'd never given individual lessons before, but I had all the certifications, so I gave it a shot. It turned out that I was overqualified. Of the two dozen or so clients who signed up, perhaps only two had any interest in spending even a second on court, let alone actually breaking a sweat. The majority were stay-at-home country club moms who'd just had their hair done and were far more interested in flirting and showing off their outfits. This was fine for a couple of weeks, but if this was what real life was like even *with* an undergraduate degree, it could wait. I once again set about scheming how I could buy myself some time and, in a perfect world, get back to South Africa. Academia seemed like my best bet.

I did some research, and learned that the University of Cape Town had a well-renowned postgraduate programme. I'd never toured the campus, but the photos looked beautiful, with grand, columned buildings and walls swathed in ivy. From the slope of the wooded mountain, it had a panoramic view across False Bay to the mainland. The catch was that I could only apply for a programme related to my undergraduate degree: international politics or law. They offered a political communication degree, which focused more on journalism and writing, so I went with that to avoid having to go near the dull, legislative stuff.

From experience, I knew this choice wouldn't go down well in central PA. When you told people you were interested/enrolled in a liberal arts degree, they looked at you as if you'd just told them you were doing a six-week course in astrology or tantric breathing. *'Writing, huh? That's not where the money is. If you were smart, you'd go learn Mandarin. Better yet, Arabic. If you were really smart, you'd learn to code, too.'* My dad had at least moved on from the egg-selling plan, and was now recommending that I get a degree in tax or, better yet, *marijuana* law: 'Lotta opportunities coming down the pipeline over the next couple years.' (He worked in neither of these fields, but I knew he liked to sound informed.)

All this advice may well have been true, but it did not answer the bigger question: why, even for those with financial stability, were our ideas around concepts like productivity still so outdated? You'd think all of this technological advancement would mean people would be allowed to work less, but no; still quantity over quality, as far as man-hours are concerned. What was the point of reaching a post-scarcity state, if we weren't allowed to enjoy it? Our idea of intelligence had been co-opted, too, defined as 'how effectively/remorselessly you were able to trade your time for money'. Thankfully, the university accepted my application, so I'd be able to defer all this for another year.

❧

I would need to find a place to live. None of the friends I'd made during my other stints still lived in the city, and there was no on-campus housing available, so my only option was to use the university's database of third-party rental properties. I found a listing calling for masters students or young professionals for a four-bedroom apartment about a mile south of the campus with a garden and a large outdoor terrace. This sounded perfect, so I reached out to the agent. She sent me a photo album of the house and mentioned there were other students interested. One spot was reserved for her son, she said, but he'd graciously offered to give the three of us first choice of the rooms, which were all different sizes and costs. I didn't want to miss my chance, so I told her to send me the lease papers so we could firm up my commitment. Within a few minutes I'd printed, signed, and faxed them back over. *The art of the deal,* I thought, thrilled to have wrapped this up so easily. *Who said landlords were greedy, entitled parasites who couldn't be trusted?*

I could now turn to the rest of the requirements which, sadly, included yet another visa. I tried to keep my chakras positive, reasoning that this time would be nothing but a formality. I'd already blazed the trail and was familiar with the consulate's shenanigans. Sure enough, I romped home ... almost.

Only one bullet point remained on the form. Its demand? The applicant (me) must not have tuberculosis. No bother; last time, I'd only needed to get a TB *skin test* (the Mantoux test was its official name). This was free, painless, and took about thirty seconds. But, as I examine the website, I notice a small

296

tweak: where it used to say 'skin test *or* X-ray', it now says only X-ray, a procedure that, in the United States, costs hundreds of dollars and is far more difficult to obtain. I call over to my friends at the embassy, who I'm positive will have fielded – even one *fucking* time – a question on this issue. But no, they protest. There have been no inquiries regarding X-rays. In fact, they assure me vehemently, this is the first time they've heard *anyone* have a question about this checklist or, for that matter, *any other* visa. And they have no memory of ever being asked about special paper stock, by the way. For good measure, they categorically deny the checklist has undergone any updates in the last few years. I thank them very much for their time.

Once I've calmed down, I can't help but be impressed by the consistency of the embassy's hiring criteria. It seems any applicants who aren't unfailingly miserable, impatient, and wholly incompetent are immediately disqualified: in dozens of phone calls to the multiple offices over the last few years, I had not yet spoken to one person who'd even *pretended* to understand my inquiries, let alone resolve them. In a way, this realization is actually heartening, as it confirms my suspicions beyond any reasonable doubt: this whole thing was a social experiment to test the limits of human compliance. For now, though, the only way out was through. I phoned my doctor to tell him that I need an X-ray.

'That can't be right,' he says, 'it doesn't make sense that they would want to expose you to radiation unnecessarily. We only do X-rays if the skin test turns up positive.'

I relay this to the consulates. My rewards are: getting hung up on by Washington, placed on indefinite hold by New York, told it is the visa office's day off by Los Angeles. Chicago stays on the line, at least, but remains steadfast in the face of logic, reason, and professional medical advice, and reaffirms the website's ruling. With this, I call my local hospital to schedule my 350-dollar X-ray.

'We don't usually let people come in for a scan without a positive skin test,' the nurse says. 'Are you sure you want to do this?'

'Not particularly.'

After receiving my unnecessary dose of radiation, I'm good to go. Once again, I set out at dawn to arrive before the visa officer's lunch/siesta. I arrive in good time and present my packet. Within a matter of seconds, he smugly turns me away, explaining that my travel insurance was in black and white. *Who could have foreseen such a thing?* I negotiate with him, knowing it will be futile.

'Oh no, maybe I can email you the document, and you can print it in colour?'

'No. We aren't allowed to give out our email addresses.'

'Well, I brought my laptop. Can I connect and print it myself?'

'We don't let anyone connect to our devices, for security reasons.'

'What should I do?'

'Well, there's a printing shop you could try nearby.'

'Oh, great.'

'But you probably won't make it back in time.'

He cannot hide his delight. I put on a dejected face and slink out of the office.

But I was only playing rope-a-dope: so non-existent was my faith the Machiavellian little cockroach that I'd brought with me a small, battery-charged generator, along with our family's printer. Once out of the office's line of sight, I sprint to my car around the corner and crawl into the back seat to set up my makeshift printing operation. After a few minutes, I stride back into the embassy; chest out, shoulders back, jaw, biceps, abs tensed, freshly printed documentation in hand. I slam them down (politely pass them through) and watch as he peruses them, searching desperately for any disqualifying criteria. But there are none. Confounded, he stamps the still-warm papers and sends me on my way.

Whether or not the South Africans were in cahoots with the special-paper people, the X-ray people, the travel-insurance people, the FBI people, or any of the other agencies or companies to which I had to fork over my money is a mystery, and will likely remain so. I thought the whole point was that we, in the West, had defeated bureaucratic bloat and meaningless authority. In the end, however, I'd not only had to figure out everything for myself, but I'd had to pay for the privilege, helping about fifteen different companies and agencies line their pockets. I'd been scanned, poked, prodded, given the run around, undermined, nickelled-and-dimed, nickelled again, kept in the dark, taunted, dimed again, scolded for not adhering to rules I'd never been informed were in place, and hung up on at least a dozen times – all for the privilege of being able to enter a country and study at one of its schools at my own expense.

But at least the hard part was over; I was heading back for round three. Older, wiser, more than ready to spend a whole year out of the country. There was nothing this place could throw at me that I hadn't seen before.

RENAISSANCE MAN

I would leave for South Africa just after New Year. Their academic year, like ours, commences in late summer, which for them, of course, means mid-January, when it's a gorgeous 75–80 degrees Fahrenheit/20–25 degrees Celsius. Even so, their patch of the Atlantic remains bitterly cold, not that this stops anyone; there exists a vibrant beach and surf culture along the coast of the country year-round.

With this in mind, the first contact I had from Fergus, the landlady's son, wasn't particularly unusual: a picture of a couple of guys in wetsuits unloading scuba equipment from the boot of a car parked beachside, with the caption, 'Dear Jack: Cape Town awaits …'

How titillating: a dynamic, adventurous spirit with a flair for the dramatic. Scuba diving in shark-infested waters? Christ, going in there with a full-blown *cage* had been too much for me. The tour de force continued with his next messages. Another future room-mate, named Lwazi, had sent a message in our group chat; his flight from Johannesburg (where he was from) would be arriving the next evening. I read on, awestruck, as the two had a brief back-and-forth in Xhosa, dialect used exclusively by the black tribes in southern Africa long before any European colonists arrived. (For reference, there aren't many white Americans who'd be able to make small talk in, say, Navajo). Switching back to English, the scuba diver now addressed me, generously offering to pick me up from the airport on the day of my arrival, explaining that he was a trainee lawyer and he had the next few days off because his team had just won a court case. For the sake of identifying each other at the airport, he told me he'd wear his most visible shirt: a bright pink commemorative T-shirt from Cape Town's Gay Pride march.

My God. I hadn't ever met a Renaissance man of this calibre: dabbling in rare and dangerous hobbies, offering himself as an ally to the long-suffering queer community, and standing in such ardent solidarity with the racially oppressed as to become (at *least*) bilingual. At this rate, he also moonlit as an underwear model who spent his free time in the Gulf of Mexico, rescuing birds and turtles that had been caught in oil spills. *How were any of us supposed to compete?*

Upon landing, I collected my suitcase and emerged into the arrivals hall, where I scanned the crowd expectantly for the six-foot-five, square-jawed behemoth (who, based on the data so far, was hung like a Clydesdale). Strangely, though, I saw no Angus stud towering above the crowd. I could see only one pink shirt, in fact, and it was worn by a short, pasty man with a … *unique* body shape. I hesitate to describe it as underdeveloped, for it appeared *perfectly* suited, if for a slightly different epoch – the Pleistocene, perhaps. He seemed confident enough in his bipedality, but his general skeletal proportions were closer to that of an orangutan matriarch, with thick, pendulous arms and bowed, stumpy legs. He was not fat, but filled out his shirt with ease; a thick layer of insulation gave his bulk a certain shapelessness. In lieu of a neck, wide shoulders served as buttresses, sloping upward and connecting to a plump, square head just below the ears (which were tiny; almost invisible from straight on).

We left the airport and killed a few minutes with scuba-diving small talk. It caught my eye that his car was an automatic – in all my time in Africa, I'd never ridden in anything but manuals. This, he explained, was actually a point of pride: none of his family had ever learned how to drive stick-shift, nor did they care to. It was preferable to just find automatic cars on the market, no matter the premium, or where they'd have to be imported from. At the subsequent lull in conversation, he asked if he could play some music.

'Please,' I said, expecting to hear fascinating, secretly recorded folk music from some tribe deep in the Kalahari. I was egregiously wide of the mark: cacophonic death metal started blasting from the car's tinny speakers. My chauffeur had to scream to make himself heard.

'Overthrust!'

I tilted my head toward the car's speaker, pursing my lips and squinting slightly to give the impression that I thought I recognized the song. Gesturing to the radio console, he went on.

'I love these guys. One of the most promising bands to come out of Botswana for years.'

We took our exit off the highway's overpass, and as we peeled down toward the road below us that would take us to our neighbourhood, he pointed out of my passenger-side window.

'That grey building there is where I get my kilts made.'

We got to Rondebosch and arrived at the house, a family home that had been subdivided into three segments; the four of us would share the top floor, while two other separate, smaller groups lived downstairs.

'My mother figured it would be easier to just buy the whole top flat while I was still in university. That way, I'd have a place to stay and could make friends.'

This was an early indication that things were to become complicated. The year-long drama would feature two main characters juggling complex roles: in no particular order, Cheryl, as landlady, 'capo', and mother. Fergus, as henchman, son, and room-mate. Their performance drew in the rest of us: myself; the waiflike, studious Lwazi; and Leslie, the tall, gregarious Englishman (who happened to be enrolled in one of my politics classes).

We couldn't appreciate its significance at the time, but in hindsight it was a microcosm of post-colonial social dynamics. In a country with a recent history of violently enforced segregation like South Africa, each social and economic stratum is acutely aware of their place in the hierarchy. Educated, wealthy (and white) citizens know *exactly* how much influence they wield, as well as how it can be leveraged. In a society so fraught with tension, the gravity of a supposed goodwill gesture thus increases in proportion to how much more social status or power one has over the recipient. As a result, these gestures are usually made more for the benefit of the giver; the social credit you can manufacture for yourself is intoxicating.

Take, for instance, Fergus, a young man in the middle of his fourth attempt at a postgraduate degree. His parents, real-estate developers who've enrolled him in expensive private schools his entire life, own a game lodge on a reserve in Botswana where they vacation as a family several times per year ('It helps us feel connected to this beautiful continent'). If he decides to learn a few sentences of the local language to commune with the natives, he is afforded papal regard (by his mother and those in his social circle, at least). This only grows when he espouses his desire for South Africa to rename itself Azania (a name proposed by many after apartheid that better represents a beautiful, diverse population and carries no baggage of colonialism). He gets bonus points for wearing a Gay Pride shirt and for hanging a Palestinian flag in his room.

The trouble arises when all this token activism fails to elicit commensurate gratitude from those whom they claim to be fighting for. In such cases, these self-proclaimed 'advocates for the downtrodden' are often the first to break character.

FENG SHUI

On paper, the hierarchy was clear, with Cheryl, the governess, presiding over four lessees of equal standing. She and her son even swore he'd signed a lease agreement, requiring monthly payments and a commitment to the same code of conduct as the rest of us. Nevertheless, Fergus sought authority in his own right, his array of theatrical gestures having proved ineffective.

I think he could sense that I wasn't going to cooperate, so he avoided direct confrontation, leaving me for his mother (of which more to come). Leslie, meanwhile, was more amenable. *His* relationship with Fergus, however, was a delicate and, indeed, highly peculiar one (more on that later, too). This left Lwazi as the only available whipping boy.

This started early on. During our first week, Fergus came into the kitchen to see Lwazi using the ice tray, which had seen too many winters and was now cracked and brittle. As Lwazi twisted it to loosen the ice, Fergus, sidled over and piped up.

'Lwazi, my friend, you must be careful. Look – you've damaged it.'

Lwazi denied this, which only provoked our host further.

'You're going to have to get us a new one. If not, my mother will find out.'

When Lwazi protested again, Fergus snapped:

'Fine, *don't* replace it. You probably couldn't have afforded it anyway.'

For those curious, this conversation happened in English. It's worth noting that, after the first exhibition in our texting group chat, none of us saw or heard any evidence of Fergus speaking in a tribal tongue ever again.

One afternoon, I arrived back from class to see a thick bundle of wires running from under Fergus's door and down the length of our narrow hallway. These were stretched diagonally across the common room to reach the internet router. I greeted Lwazi, who was in the kitchen. He was unsupervised, so I knew something was up. I followed the wires back to their source. I knocked, and when Fergus came out, I asked him what was going on.

'Well,' he said, 'the Wi-Fi didn't reach down the hall before, and I need the internet to be able to play my games. So, I had my mother send someone in to set me up with a hard-wire connection. Much faster.'

But the wires would be a hazard, I said. We'd have to tape them to the ceiling or at least along the skirting. He disagreed.

'I'd really prefer we left them as is. I don't want anything to slow the connection.'

Lwazi, whose room was even farther from the router than Fergus's, chimed in.

'I've been emailing your mother since we moved in, asking her for a new router or a wire like yours. I pay her for Wi-Fi like everyone else, and I can't even connect to it.'

'Shame.' Fergus said, grimacing. 'You'll have to take that up with her.'

With this, Fergus realized the apartment itself offered a wealth of opportunities to assert himself. If we so much as tried to shift the furniture in the living room or on the terrace by a few inches, he would object, reminding us that we couldn't take any action without first asking his mother. He didn't make the rules, he said apologetically, his hands were tied. Whenever we sought his mother's assent, of course, she deferred back to her son.

The main corridor, which ran from the front door down to the bathroom and the outdoor terrace, was lined with our bedroom doors on the right side, while opposite was the lone, wide entrance to the common room, whose right-hand wall had an identical opening to the small galley kitchen. Despite the property having been listed as 'fully furnished' in the rental housing database (thereby allowing the landlady to charge a premium for such a luxury), my room-mates and I had arrived to find our rooms almost completely empty, save for our cheap chipboard desks. Our flat's only 'fully furnished' room was the common room, which contained a jumbled assortment, including a couple of old rattan dining chairs and a long wooden table that was positioned immediately inside the room and parallel to the hallway, meaning it barricaded almost the entirety of the room's entrance. Since it seemed obvious this room would be a thoroughfare,

I wondered aloud whether the table must have been in limbo, awaiting a group decision. Fergus informed me that this was not the case.

'This was intentional. I wanted to make the entrance a little smaller so that the room would feel more cozy. Anyway, we have to make it work, because it can't be removed from the apartment – the stairway's too narrow. The workmen had to winch it up over the balcony, like you would with a piano. They did the same for the bed.'

'Wait a second. You got a *bed?*'

'No, I'm not talking about the bed in my room. The workers managed to get *that* up the stairs without much trouble. No, *this* one.'

He patted the piece behind me fondly. It looked kind of like a futon, but one without the capability of transforming into a couch. As such, it was just a low wooden frame supporting a lumpy, queen-size mattress upholstered in brown velour corduroy. Given its position in the centre of the room, it seemed to have pride of place.

'This old girl,' Fergus noted with a smirk, 'has gotten some *heavy usage.*'

'Oh, yeah?'

'*Oh*, yeah. She's seen some terrible, terrible things … My room-mates and I used to go *apeshit* at our last house. This bed was in the living room, and I would crash on it all the time. If I passed out on the floor instead, Jess or Louise would use it to hook up with the guys they were dating. Great memories.' He fell into a misty-eyed reverie, idly tracing one of the cushion's larger stains with his fingertips.

It was unclear whether those girls knew the extent of Fergus's familiarity with their intimate affairs. For the record, there's no shame in wing-manning for a friend, but there was something slightly pathetic about always being this *particular* bridesmaid. I asked him if he'd ever had the chance to use the mat with a partner of his own.

'No, but don't you worry. Let's just say I've gotten my fair share of kicks on here.'

He smirked again, and fell back into another daydream, before eventually continuing.

'Anyway, she's retired now. But she'll be great for our movie nights.' He gestured to the screen in the corner. 'I had the construction crew set that up. It took them a few hours with all the drilling.'

Not only had he chosen the small corner between the room's two adjacent openings, but rather than angling it to face outward into the room, he'd instructed the crew to mount the screen flush to the wall, facing lengthwise down the table, making that the only possible vantage point. But anyone seated at the head closest to the television, would have been facing directly away from the screen, while the other end was pushed up against the opposite wall, which left no space for a viewer. As for the longer sides, sitting on one of these would have left you completely in the hallway, leaving just its opposite side as a workable option. Even then, it wasn't clear what our host envisaged would happen when more than one person wanted to watch something. He'd had mounted the screen at waist height, which meant your only hope would be if your guests were happy to sit from shortest to tallest.

The table remained a point of contention throughout the year. As we were all on different schedules, our meals were usually prepared and eaten alone while watching television or, weather permitting, out on the terrace; the table was thus wholly unnecessary. Fergus, for whom it represented community and group bonding, was crestfallen. He told us he was disappointed we didn't appreciate the table and the trouble he (i.e. the team of labourers) had taken to transport it: 'It's just that, at my last house, the girls used to prepare big, delicious meals that we'd sit down and eat as a group. I was thinking we were going to do the same thing here.' But I'd never heard him express any such wishes in the months prior. What he *had* done was bring back large Tupperware containers of pre-prepared curries and pastas after his visits to his mother – not that a share of these was ever offered to any of us. As such, the evidence at hand indicated the actual state of affairs was something closer to: 'At mealtimes, I used to hang around my room-mates, who were already intimidated by me because of my exhibitionism and my penchant for falling over and passing out near (or under) where they were hooking up with their boyfriends, and because they knew my mom owned *that* flat, too.'

❧

Electricity was another sticking point, not only in our flat, but the country at large. The power grid, nationalized and under constant changes of (mis)management, was nowhere near as productive or efficient as it needed to be. This meant there was an official rota dictating when electricity would be rationed (cut) in each area. The government euphemism for this was 'load shedding', as in: we (you) must lighten the load on the poor old power grid. Depending on the severity of the shortage, a district might be without power for two hours once or twice a week, or every day for all but a couple of hours. Given that we usually had forewarning, this situation wasn't too bad, but it was irritating being stuck waiting for someone to finish using the kitchen when you hadn't yet made dinner and the power was about to disappear until three o'clock in the morning.

Although electricity was supposed to be included in our rent, Cheryl only gave the flat about thirty dollars' worth per month. When this ran out, we had to go to the grocery store for a top-up voucher. The three of us took turns making these trips all year, as we could never quite make the main ration last the month. Our appliances, made to European standards, had no sympathy for South Africa's nonsense. Putting one load of clothes through the German-made washing machine, for example, drained about one-tenth of our monthly supply. Manageable, you might have thought, but this meant that Leslie's habit of never wearing anything more than once before putting it through a full-wash cycle quickly became exasperating.

Warmth also became a precious commodity: many houses in South Africa aren't built with insulation in mind and are drafty and frigid throughout the sodden, stormy winters, during which overnight temperatures drop close to freezing. With no central heating or fireplace, the only option was to get an energy-guzzling space heater to warm your room.

Not that Fergus was concerned with any of this. I arrived home one afternoon to see him standing in the kitchen over two workmen who had shifted our refrigerator, stove, and washing machine to make space for a dishwasher they were installing. I tried to figure out what was going on:

'What's going on?'

'Well, I hate doing the dishes, so I'd rather just use the machine.'

I believed him: our kitchen had a normal sink but also a larger, industrial one in the back corner; though his visits to the apartment were short, he would fill these with overflowing piles of used pots, pans, and tableware before leaving the house, knowing that the cleaner would take care of them (most middle-class homes, and student accommodations, had cleaners in South Africa).

'Who's paying for this?'

'Don't worry, my mother bought it for me.' I asked if he or his mom would provide more electricity, given that his use of the dishwasher would affect the budgets of the rest of us.

'Well, no, since you'll all use it, too!'

'No, we won't. We should have taken a vote on the decision beforehand.'

He called off the workmen, but the machine sat in the kitchen for weeks before his mother had it collected. It was like the Humvees and other vehicles left behind by the US across the Middle East: you may have won *for now*, but don't forget who's roof you're under.

❧

Shortly afterward, we received an invitation to one of Fergus's (self-professedly) legendary get-togethers. I assumed, at first, that this was in the spirit of reconciliation after the dishwasher dispute, before noticing the party's location was listed as our apartment – he'd had no choice but to invite all three of us.

The first paragraph (of seven) began by stating that, 'Fergus and The Bhoys have become fatigued from academic rigor. In light of this, we've decided to inject some levity – by which we mean alcohol – into our work process by having a night of intoxicated presentations.'

The rules for participation were then laid out: upon arrival, all entrants were required to chug two pints of beer. When your name was drawn to present your academic thesis to the group, you'd need to take three shots before getting up to speak. Everyone had to gulp down *another* pint before each subsequent presentation, and *another* upon completion of the final one. There was also an appendix detailing a drinking punishment featuring scuba equipment: a full-face dive mask was strapped to the kneeling victim's face, while assistants (standing) poured alcohol down the mask's air tubes from above. The only way for the victim to avoid drowning in a litre of mead or boxed wine was to try and swallow the deluge before it filled up the mask.

I have no idea how many attendees there were in the end, but when I checked the online invite page on the evening of the party, I saw it had been sent to fifty people. Seven had RSVP'd that they'd be attending, with the rest having declined the invite (in fairness, a handful of that forty-three had said they were 'maybe interested'). Of the seven who had confirmed, Fergus was one, while three more were us, his cohabitants, who had confirmed our attendance out of pity; on the day he sent the invitation, he'd visited each of our rooms in turn to stand over us while we read the whole thing, lingering in our doorways with a self-satisfied smile until we promised to attend. Legendary as the party promised to be, all three of us made plans to be out of town that weekend. The fact that none of us would be there worried me.

As expected, I came back on the Sunday afternoon to find our apartment in tatters. Furniture and cushions were strewn through the hall and common room, and there was a mysterious, sticky coating on most of the hardwood floor. Outside on the terrace, our small barbeque grill had been tipped over, evidently while in use; its spilt coals had burned a large hole through the artificial turf which covered half of the porch. Knocked over onto the cement a few feet away were the chairs from our dining room. For convenience, someone had brought the entire cutlery drawer out from our kitchen, but it now lay upside down, its contents scattered around the terrace. It had rained heavily in the early hours of the morning, meaning all of this debris, which had already occupied varying states of decay and shabbiness, was now completely waterlogged.

One dining chair, however, had suffered no water damage, having been left inside. But it had only been spared because someone had stood on it with a peg-leg, puncturing the brittle wicker and leaving a gaping hole. Noticing that it looked exactly like the chair in which Daniel Craig's James Bond gets tortured in *Casino Royale*, it crossed my mind that perhaps a similar ritual had taken place at Fergus's behest. Perhaps the threat of emasculation had been nestled somewhere in the web of subparagraphs and clauses dictating the party's rules:

- Rule #4: If you pass out with your shoes on, you're fair game to be meddled with: you might have your hand soaked in warm water to make you pee yourself, your eyebrows shaved off, or a Sharpie moustache drawn on you – who knows!
- Rule #5: Don't change the music without asking one of the hosts first!
- Rule #6: In this house, we hold our liquor! Anyone caught (or suspected of) vomiting will be stripped down and tied to our favourite chair, which we call The Castrator. A mysterious, disfigured European

man will then arrive to bludgeon your penis and testicles with a rope flail.

- Rule #7: *Party on!*

Unlikely, but I resolved to keep an eye out for anyone whose walk of shame looked particularly ginger.

The host of this bacchanal was AWOL for the next seventy-two hours. He eventually reached out via text message to explain. His mother had invited him home for a lunch she was preparing, you see, and he hadn't wanted to disappoint her. He assured us, however, there was nothing to worry about: the cleaner would be there in the morning to tackle the spillage, which had already stained and begun to warp the floorboards.

ECCE HOMO
(... OR AT LEAST BI-CURIOUS)

Despite his avowed pride in our apartment's Feng Shui, Fergus mostly stayed at his parents' house nearby. On the occasions he stuck around a little longer, we'd know at dawn, when his alarm clock, set to play screeching death metal, filled the house. On several occasions, we got Who Let the Dogs Out, or his other favourite song: a high-pitched, sped-up techno remix of [I'm a] Barbie Girl. The music would usually fail to wake him and would continue playing until one of us could take it no longer and sprinted down the hall to thump on his door, whereupon he'd hit snooze and fall back asleep for another few hours. At some point in the early afternoon, he'd lumber down the hallway to take a shower (during which the music would resume), before donning his skirt and splaying himself out in the living room. Not his kilt, his skirt. These were two separate things. The first weekend after we'd all settled into the apartment, I was awoken by Overthrust from the bathroom, to which he was screaming along while he showered. I went into the kitchen to make a cup of tea. He must have heard me pottering around and stopped in to greet me. He was half naked, with a towel around his waist. I mentioned that his music had carried through the whole house and may have disturbed the others. He was sceptical:

'That can't be. I tested all the rooms for sound.'

Of that I have no doubt, I thought, and gave him a wide berth on my way out. He'd begun shifting the bed toward the television. For the next couple of hours, I heard Japanese-sounding shrieks, wails, and moans from the common room. I tried to wait him out, but after a while I got too hungry and decided to risk venturing out again. I saw that the noises weren't coming from him, but from the anime he was watching at maximum volume. But this was the extent of the good news.

Our host lay in repose on the mat. With no smaller pillows or cushions at hand, he'd interlocked his hands behind his head to tilt it up and create a sight line for his low-set eyes to peer over his bloated gut. His legs were bent and spread wide, as if he'd been told to assume the position for a Pap smear. The towel he'd been wearing earlier had been replaced by a low-rise Indian sarong, the hem of which was hoicked up around his furry haunches. Had I been standing anywhere inside the room I'd have easily seen the full contents of his silk-swaddled cornucopia. I could tell he was hoping I would compliment his fashion choice. When I obliged, his eyes lit up.

'Thanks! My room-mates used to hate me wearing this around, but it's so airy and light, it just lets everything stay dry all day. It's my favourite thing to do – relax for a while and enjoy the breeze. I like to treat myself to wearing it on the weekends. Thank God, it's a guys' house now, so it won't be an issue.'

He was correct to presume that we'd likely forgive a bit of public indecency (boys will be boys, and all), but it soon became obvious that he was getting a little too comfortable; we quickly cottoned on to our blue-balled elder's fantasy of being caught lube-handed. Almost every weekend was marked by one of us walking in on him groping, petting, or kneading himself.

If it hadn't already been apparent, we had a serious chain-of-command issue on our hands, equivalent to a prisoner waking up in his cell to find the warden fondling himself in the opposite bunk. The man whose silk skirt we were looking up didn't just have *himself* by the balls; he had a bedroom just down the hall in a foreign country miles from home, as well as all of our rental deposits.

Our only recourse, it seemed, was to look on the bright side. With his Neolithic physiognomy and bone structure, this genetic throwback had evoked for me, and now *us* (lucky you!), a vivid tableau. A portal had been conjured through which we could witness a crucial chapter in human prehistory: the first instance of early man dabbling in auto-eroticism.

Ecce (pre-)Homo: left alone in his tribe's grotto, a juvenile Australopithecus, the damp walls around him illuminated by nothing more than flickering torchlight. He can't sleep, lounging on the ragged communal bed-mat, restless, bored. He peers down at himself, tracing his fingertips along his stout, hairy trunk before pulling his mammoth-hide briefs aside ...

❧

These incidents became increasingly charged. Leslie had been open about his sexuality from the beginning, meeting people and using dating apps. Fergus, in fairness, was also candid, albeit markedly less successful, regaling us with his stories of bad luck with women. None of this was surprising. As if his physique wasn't enough of a hindrance, his looks pushed things over the edge. The entire top half of his face was dominated by a flat forehead, confining his features to the lower half. Wide-set, beady eyes sat on either side of a round, pink nose; their pupils (slightly misaligned) floated in dull, watery-grey irises. Meanwhile, a pelt of dense, mousy hair crawled up from behind his collar, spreading up and over the back of his head, becoming wispier as it ascended. All this to say: I knew that his odds of finding a living, female *Homo erectus* were slim to none. My sense was that his only hope was to re-consider his preferences. When the odds are stacked against you, you can't afford to discount half of the global population.

Not long into the year, he reached the same conclusion. His long-standing desperation was the kindling, and Leslie was the spark. Within a couple of weeks, it became clear the Englishman had been chosen as the boy-king's 'court favourite' (à la Renaissance Europe). This arrangement was, for a time, mutually beneficial. Fergus had a confidant and muse on whom he could shower gifts and flirtatious attention, while Leslie got to enjoy a charmed life around the house, keeping the older man at arm's length but – and this part was crucial – *never* totally rebuffing him. I warned Leslie that keeping his suitor's hopes alive would only stoke the fires of lust. Yet, despite constantly complaining that Fergus had no respect for his boundaries, it seemed Leslie was happy enough to role-play as geisha. The doors to his bedroom served this analogy well: his room, sandwiched between mine and Fergus's, was directly across from the common room and, inexplicably, had large glass French doors covered with a semi-opaque film, which failed to provide occupants with even the slightest hint of privacy if the room's lights were turned on. For a passersby or anyone in the common room entrance, it was like a shadow puppet show – a happy coincidence for any voyeurs in the house. From his place on the communal sex bed, Fergus could monitor Leslie's comings and goings, and would knock on the door whenever he saw illumination and wanted company.

ૐ

Part-way into the year, Fergus organized a little soirée. The other invitees were a motley crew made up mostly of his classmates, one of whom was a dreadlocked Zimbabwean who quickly befriended Leslie. Toward the end of the night, we noticed the pair had absconded from the terrace. When we saw Leslie the next day, he complained that the shirt he'd been wearing during the party had disappeared, and asked Fergus for the guy's number to track the item down. This struck a nerve:

'I'm not giving you anything until you tell me what you did with him. He's my friend, so I have a right to know. And don't you dare lie. I saw you holding his hand as you led him into the apartment.'

He tried to deliver this ultimatum calmly, but we could hear his voice trembling. Leslie, non-confrontational to a fault, confessed immediately.

'Well, we had a few drinks downstairs, and we were flirting a little. But when we went up to my room, he told me he had never hooked up with another guy before and wanted to take things slowly. So, we just kissed for a little and then fell asleep.'

This innocuous-sounding account was apparently unsatisfactory: the blood drained from Fergus's face, and he stood up abruptly. 'Ha! I knew it. Leave it to *you* to turn the gorgeous black guy.' He stomped down the hallway and into his room, slamming the door behind him. (He never did give Leslie the number.)

After a few months (and seven or eight more attempts at seduction), Fergus realized that being passive wasn't going to cut it. Picking a weekend when neither Lwazi nor I were home, he made his move, throwing together a last-minute date night of sorts. Needless to say, the Zimbabwean shirt thief was not invited.

Upon returning to the house the next morning, I heard a muffled sobbing coming from down the hallway. Through the frosted glass, I saw a Leslie-shaped shadow sitting on his bed. When I knocked, he invited me inside. He looked distraught and completely defeated. I asked what on earth had happened.

As soon as I'd left, he explained, Fergus had shown up and settled in, taking a long, steamy shower before slipping into something more comfortable – a new, chiffon sarong. Of course, this alone was no longer noteworthy, nor was the next revelation, that Fergus had invited himself into Leslie's room and sat on the bed, venting his frustrations about how he just *couldn't* seem to find love. I imagine it was something like: *'People are too scared to take risks these days, don't you think? Sometimes you've just gotta take that leap of faith ... Your true love might be right in front of you, but you're too scared to open your eyes ... But hey, what do I know? I'm just a fool, a hopeless romantic.'*

Fergus eventually left the room, but could still be heard prowling through the house, so Leslie stayed in bed to avoid him. After an hour or so, the footsteps subsided, and Leslie ordered a dinner delivery. When he left his room to go meet the delivery driver outside, however, he was ambushed: lying on the sex bed, fully exposed, was Fergus, his cock 'fluffed' and at the ready, its drooling tip poking out from a dense tangle of pubis. (Evocative, if you will, of a wrinkly, foreskinn'd quail's egg presented, with aplomb, atop a woolly cushion.) Seeing Leslie, he sat up with a start.

'Oh, shit! I didn't think you were home!'

'But you've been stomping around the house all day – you would have known if I'd left.'

'Hmm. Maybe. Anyway, you're lucky you didn't come out. Whenever I'm home alone, I take all my clothes off ... I feel much freer that way.'

Unable to summon the hutzpah to end this debacle then and there, Leslie went down and collected his food, then let Fergus follow him to his room and flop down on the bed. 'What did you order? I'm starving.'

At this point in the retelling, Leslie burst into tears.

'He just kept eating my chicken nuggets. He wouldn't stop. He didn't even offer to pay me back.' (To me, this sounded like exactly the type of 'communal' meal our host had been hoping for.) Leaving Fergus on the bed behind him, the distraught Leslie then ran out of the house. The golden handcuffs, it seemed, had finally grown too tight for comfort.

The nugget-thief's scheme having failed disastrously, he retreated to his parents' house to regroup. It was weeks before we saw him again.

❦

During our winter break in June, I flew up to England to visit family. A few days before my flight back to South Africa, I woke up to multiple panicked messages from Leslie, reporting that there had been a break-in. A large glass pane in our front door had been smashed and its retractable security gate was broken, having been forcibly wrenched out of the way to access the door behind. He'd discovered the damage the morning after staying at a friend's house; fearing he'd stumbled into an active crime scene, he decided to make himself scarce. He was now sitting at a café and waiting to hear from Fergus, whom he'd seen the night before but was now nowhere to be found.

Leslie's story was worrying, but entirely believable. He'd been robbed at knifepoint outside a nearby restaurant a month before. Then, just a few days afterward, as he and I sat on the terrace after class one afternoon, we'd heard screaming coming from the nature trail behind our house. A woman had been mugged in broad daylight, with the offenders running off down the trail. Worse *still*, one of the flats downstairs had been ransacked just a week or so before I'd left; the investigation was still ongoing. After a few hours of detective work, we discovered that the architects of *this* break-in attempt had been associates of none other than our resident scuba and hentai connoisseur.

One of his old room-mates had organized a night on the town to introduce him to her new boyfriend. At Fergus's insistence, Leslie had joined the group on the first stop of their bar crawl but, as it turned out, the girl's new boy-toy was a hardline Afrikaner nationalist who soon became verbally abusive and homophobic toward several strangers at the bar. Leslie felt uncomfortable and made a plan to split off to join his own friends elsewhere in the city. The honeymooners caroused for another few hours before deciding to send their evening hurtling toward its climax. Having come into town from Stellenbosch, they'd been invited to spend the night at our apartment by Fergus, who was meeting a different friend at a death-metal nightclub and wouldn't be home for another couple of hours. After returning from the bar, the couple realized that they only had *one* of the three necessary keys to get into the flat. They were able to make it into the house but were locked out of our flat's front door and its metal security gate. The girl had her smooth-brained associate force open the gate to gain access to the door, then smash one of its glass panes to get access to the handle. What our Bonnie and Clyde didn't realize was that it was a deadbolt lock – turning the handle without a key wouldn't do anything. Not that they didn't still try *a* key: they wedged one of their own into the lock, which naturally got stuck. Having broken the glass with his bare hands, the racist lunkhead now sported a deep laceration down his arm and was haemorrhaging blood. Given that our foyer lacked the materials for a tourniquet, the duo decided their best course of action was to drive back to the farmlands from whence they'd come.

We dutifully reported the damage to our landlady, requesting new locks and keys for the gate and the door. Ideally, we noted, these jobs would be completed with haste, given the break-in downstairs just a few weeks earlier.

'Not to worry,' she told me over the phone, she understood our concerns and had already set the wheels in motion. It was her son who'd been tasked with this crucial job – and whom I could now hear whining in the background. His handling of the issue was predictable: although we could enter our apartment, it was almost a month before any repairs were made. Even then, it was *another* couple of weeks before we got our own keys; he had only made one for the three of us to share. The delay, he explained, was because he'd been waiting for his monthly allowance (he'd blown last month's on the first day, buying an action figure that had just been released). In the meantime, he decided that he'd be more protected – both from crime and our judgment – at his parents' house. For those of us without the luxury of retreating to country estates, this period was nerve-wracking.

Once we had new keys secured, we could be open with our disdain. Toward the end of the year, this meant that actual interactions were rare and Leslie, Lwazi and I relished these spells with the cessation of sexual loitering, intrusions into personal space, and death metal shaking the walls of the house at all hours. Occasionally, we'd hear comings and goings from his room in the middle of the night, or we'd get back from class to see Fergus rush to his car and crunch his way through a three-point turn before careening away. It became obvious that he was only using the apartment for its washing machine and commode; we'd often come out in the morning to see the cleaner desperately plunging a toilet he'd blocked. The most we heard from him were reminders on his way out the door: 'The next time one of you goes shopping, could you grab some toilet paper? I would, but I'm in a rush. Oh, and please make sure it's quilted. That one-ply stuff is too thin. I don't enjoy touching my asshole every time I use the toilet.' Obviously, we had ample reason to doubt this, but splurged for the two-ply anyway so that we didn't have to revisit the topic.

One night, I accidentally postponed one of his stealth bombing runs. A couple of classmates and I were outside on the terrace having a beer. We'd presumed Fergus wasn't home, as there had been no light visible through his window. I needed to pee and ran inside to the bathroom. The door creaked open behind me.

'Gimme a second and she's all yours,' I said, without turning around.

A low, hoarse grunt came in response. By the time I finished, the hallway was empty. Though we looked outside and saw his car, there was still no response when we knocked on his bedroom door. The next morning, I awoke to find the toilet spilling over, clogged by a mammoth, gruesome bowel movement. His car was nowhere to be seen.

POLITICAL
(MIS)COMMUNICATION

Even while Fergus was away, I had little faith that we were ever *truly* alone. These suspicions weren't entirely unfounded: I came back from class one afternoon to find a crew of boiler-suited African workmen climbing up into my floor-to-ceiling closet. This, they claimed, was the only access point for the flat's attic; they'd been commissioned by Cheryl to check on some electrical fittings. It wasn't safe for me to be in the room, since live wires were being brought in and out, so I waited nervously out on the terrace. After an hour or so, they folded up their aluminium ladders and departed. I ransacked my desk and cupboards but could find no evidence of the cam(s) I was sure had been installed.

As you can imagine, the constant threat of surveillance put a damper on our social lives. Although the apartment and its large terrace would have made for an ideal place to hang out, we were hesitant to invite any friends over, let alone potential love interests. Not that I had many leads on the latter front. As far as classmates, there were only a couple of options. One of these was a redhead from one of my politics classes. She seemed friendly, but it was never going to work out: she was Jewish.

Perhaps I should clarify. Her religion wasn't the deal-breaker ... but it *was* technically the reason I never got to make a move. It was a couple of months into the school year, and the South African election was around the corner, so our politics professor brought up a few other current elections in the Global North. Jeremy Corbyn's (an English party leader who'd spoken out against Israel's occupation of Palestine) name came up, and when another student professed her support, our redhead (Hannah, we'll call her) huffed, and began to type into her laptop furiously. When the professor agreed with the other girl, Hannah slammed the computer shut and began to pack her things. When he noticed and asked if everything was OK, she said, 'You know *full well* that I'm treasurer and co-chair of the school's Jewish league. How can you openly endorse a man who would kill me if given the chance?'

She withdrew from the class, and I never saw her again. A shame, because I knew just how to get into her good books: mention that my country, the US, had given her people in Jerusalem a quarter of a trillion dollars since World War II.

Notwithstanding that, I couldn't help but feel that her anger was misplaced: there were politicians far closer to home who had made explicit threats of violence. South Africa's Julius Malema, the iconoclastic leader of the Economic Freedom Fighters (EFF), a far-left revolutionary party whose platform included policies like the seizure, if necessary by force, of white-owned land. He'd been in the headlines since the beginning of that year, as there was a general election in May. His demagoguery appealed to the country's significant swath of angry, young, and unemployed black men. He'd been accused of hate speech and inciting racial violence with his charged rhetoric, and even with his campaign's theme song, called Shoot the Boer (Boers were the country's white Afrikaner farmers). He allayed concerns by reassuring the press that: 'I am not calling for white genocide … At least not yet.' *Phew! Well, that's a relief, Julius, thanks for clearing that up.* As far as I was aware, Corbyn had never worked his crowd into a lather with chants of 'A bullet for every Jew,' but what did I know?

Malema had actually been scheduled to give a speech at the university just before the election, but this was called off after he became embroiled in another scandal. While we were discussing all this at class the day afterward, I was surprised to hear Leslie disagree with the teacher, a black man, as he was telling us about Malema's questionable background – the politician was suspiciously cozy with the ANC and had faced charges of corruption himself. The teacher suggested that Malema hadn't always been so inflammatory and explicitly racist; perhaps his histrionics were a deliberate tactic (perhaps with the backing of the ANC) to create division and undermine the credibility of left-wing politics in the eyes of the general public. Leslie interjected.

'I don't think he's racist.'

'What do you mean?'

'Well, that would be impossible.'

The professor looked confused. 'What do you mean?'

'It's only racism if you punch downward.'

'So, what would you call it?'

'I dunno. Racial discrimination, maybe.'

I was fascinated. This was the sort of interaction you only ever saw on the internet: a white person (who enjoyed saying things like, 'We [white people] have had our chance to speak – it's time to shut up and listen.') educating a black person as to what was and what wasn't racism.

One problem with these sorts of arguments is that they can be just as paternalistic. For example: as it stands, apartheid remains the socially accepted justification for the country's malaise. But let's say that a decade passes, and that AIDS, poverty, and unemployment rates are all still close to 50 percent. What about *five* decades? When would we confront the possibility that the well has been poisoned? That the damage wrought by the post-apartheid cocktail of unpreparedness, ineptitude, corruption, and rent-seeking might be irreparable? There have been dozens of highly educated people in positions of power in South Africa, all of whom should have 'known better', by *any* Western standard of ethics or law.

The trouble, I think, is that we're taught that other races and ethnicities are ideological and emotional monoliths: if we've wronged one of 'them', we've therefore wronged all of 'them', and thus owe each of them equal supplication. Similarly, we can't imagine that a historically oppressed or disenfranchised group could also be capable of perpetrating wrongs of its own.

This means that, in our frantic desire to show solidarity, we're liable to back the wrong horse: if I were Malema, I'd be *thrilled* to see white people convincing themselves (and each other) that they can't be the victims of racist attacks. It's rhetoric like his, of course, that has contributed to the growing fear felt by many Afrikaners, thousands of whom have fled inland to their private colonies, given their lack of faith in the government and law enforcement.

We might be tempted to declare, with a sneer, that it's good for those racist hillbillies to get a taste of their own medicine. It's all very easy for us to espouse our virtuous-sounding ideals when our lives will not change one way or the other. Indeed, think of how happy I was to defend Ravesh when it wasn't my apartment he was set to move into. It's similar to how every perfectly manicured front lawn in those gated, cookie-cutter neighbourhoods back in the northeastern United States had 'Refugees and Immigrants Welcome' signs. But in a country like South Africa where tens of thousands of people are murdered every year, and when racially motivated attacks on white farmers (and students, as we learned in Stellenbosch) *do* happen, perhaps we can't blame folks for battening down the hatches.

This speaks to the real issue with basing our politics (and, in many cases, our very personalities) on superficial attributes: it keeps us divided and distrustful of each other, and this is no accident; individualized resentment (scolding and tattling on each other) will not and, by definition, *cannot* threaten entrenched power.

It's for this reason, by the way, that I suspect the spectre of apartheid (and other identity-based conflicts) will be allowed to linger. If we were to conclude that the millions upon millions of suffering, diseased, and dead are no longer casualties of apartheid, this would open the door for a far more odious corollary: that they might be dying because of an economic system that incentivizes – and indeed *demands* – the dehumanization and exploitation of billions of people, irrespective of their race, ethnicity, or creed.

৵

To the West's (and the Boers') relief, I'm sure, the election was not won by Malema's EFF, but by the incumbent ANC.

The party was no longer headed by Zuma, but Cyril Ramaphosa. He had a comparatively better track record than Zuma, but he too had questions to answer, having allegedly been involved with the horrific Marikana massacre in 2012, in which thirty-four striking mineworkers were killed by police. All charges against him were eventually dropped, so all's well that ends well, I suppose. (All ended *especially* well for his friends and family, who were all gifted various executive jobs in both the government and industrial sector, including Eskom, owners of the country's spotty power supply.)

Ramaphosa had brought with him a professed desire to crack down on government malfeasance and corruption, but he was just another careerist who'd made his millions in the post-apartheid privatization boom. Would he really have the inclination, let alone the testicular fortitude, to tear down the very structures that had allowed him to reach the top? If he was the best they could do, I was surprised how low the bar had fallen. All of these ANC bigwigs had been imprisoned (or worse) for their anti-apartheid activism in the decades before the regime fell. That would have taken immeasurable bravery, resilience, and cooperation. Today, though, the reputation of every single one of them, even Nelson Mandela's ex-wife, Winnie, was clouded by a raft of corruption and embezzlement charges, civil and criminal lawsuits, and more. If *their* morals could be bought, no wonder voters were tuning out.

The ANC's latest victory may have bought them a few years, but it was clear that unrest, fomented by parties like the EFF, wasn't going away. The West, I knew, would be watching on anxiously. It would be a nuisance if a pro-worker, anti-capitalist, anti-imperialist revolutionary party gained power. Especially as the golden years of total impunity for interference by the CIA and FBI (as evidenced in regions like Latin America, the Congo, Iran) may have passed.

Of course, it's not like South Africa's elites would cede power willingly. To ward off the guillotines, however, they might consider following countries like Angola, Egypt, and Ethiopia, and open their doors to investment by China. In our estimation, doing this would put South Africa at risk of becoming yet another Chinese vassal state. Our hackles were already up: within the last couple of years, China had established a military base in Djibouti, further strengthening its foothold on the continent. *Just what were they playing at, the Chinese? Who d'they think they are … Us?* The last thing we want is for the news to get out that, contrary to our proclamations, liberal democracy might *not* be a prerequisite for capitalism – which, to our dismay, was already being proven by our Chinese friends, who looked to be holding their own just fine.

ॐ

That's the fun thing about being the US, or its faithful sidekick, the UK: we get to offer other countries all the unsolicited geopolitical advice we want, labelling them as failed states, plutocracies, oligarchies and scolding them for corruption, cronyism, and racketeering, while being immune to all that. That's what we were taught, at least. But it seemed like that this façade was starting to crumble.

Indeed, I couldn't judge anyone in South Africa for losing faith or interest in their political system, as the same thing had happened to me, and many others, back home in the states. Faith in every branch of government has been on the decline in the US for years. *Hundreds* of millions didn't even voted in 2016; Trump won that election with the support of only a quarter of Americans of voting age. By my third visit to South Africa in 2019, the campaign and media cycle for the 2020 presidential election was fully underway. And what lessons had we learned after Hillary Clinton's defeat three years earlier? The best we could do, apparently, was Joe Biden, who could barely walk and was apparently losing his mental faculties by the day. The combined age of Biden and Trump was closer to 200 than 100. Were either of them truly in touch with the struggles faced by their own citizens, let alone with the digital, modern world, and the challenges it presented?

People had begun to realize that, in practice, our political and economic elites weren't fundamentally different: regardless of political party, they all navigated the same social circles, summered on the same islands, and had the same private equity, legal, and public relations firms on retainer. Their main priorities – namely, the consolidation of power – were aligned. The disdain with which working people were regarded was clear: in terms of wages, the working class was no better off than it had been half a century ago.

Our most salient divide, it seemed, was not between the orthodox parties, races, or genders. It was between those who believed, or claimed to believe, that the systems around us were fundamentally just, meritocratic, and fit for purpose, versus those who did not.

There was *one* traditional demographic that captured this: age. That proponents of the status quo skewed older was unsurprising. A system built on self-interest will, necessarily, tend toward concentration of wealth rather than dispersion; sure enough, 'baby boomers' are a staggering 90 percent wealthier than those half their age. Systemic change would not only threaten their nest egg, but their very identity, their sense of cultural and moral superiority. In fairness, these folks grew up during the post-World War II economic boom, a period of unchecked military and economic hegemony – there was no reason not to buy into the American dream. Of course, a system based on schoolyard claims like 'might makes right' and 'finders keepers' is good fun if you arrive early enough to get in on the looting. Although this system is supposedly fueled by work ethic, private property, individualism, etc., it was far more effective when we also had slaves and far-flung colonies that kept things running in the background. (We hear much less about all that stuff.)

Anyway, if all that was the metaphorical carrot, we'd be remiss not to acknowledge the stick: these folks also grew up being instilled not just with the fear of God, but of nuclear war with the communists, and were therefore perpetually on the lookout for Soviet sympathizers, double agents, KGB apparatchiks, etc. As far as millions of people are concerned, it's still the Red Scare. James Bond's gun-slinging heroics from those opening GoldenEye levels is their wet dream.

With all this in mind, it's no surprise that our cultural and political discourse doesn't engender substantive, good-faith discussion; its primary concern is nipping existential threats in the bud. Luckily, this isn't hard to do. Since nothing – art, discourse, innovation, and even sport – can exist outside our economic system, any dissent is seen as brazenly ideological, radical, sacrilegious. The status quo, on the other hand, is not rooted in *any* ideology – other than common sense, of course. This saves us a lot of time. We can skip straight to ad hominem; calling into question our opponent's sanity, intelligence, and/or moral fibre with accusations of things like hypocrisy, laziness, and naïveté. These diversionary tactics allow heinous, inhumane acts to be perpetrated in plain sight, immune to criticism on the basis that 'this is simply the way things have to be'.

If I thought the political landscape was bleak, the dating market wasn't proving much better.

READY FOR
SOMETHING REAL

With our ginger Zionist having ruled herself out, this left Jasmine, an Indian girl from my second-semester film class. She was pretty; thin, sporty, nice hair. The fly in the ointment was that she was almost as tall as I was. I'm a big, strapping boy, and I want to feel like one, dammit. But, with other prospects thin on the ground, I decided to risk feeling emasculated.

We swapped numbers after class one day, and I messaged her later on to suggest meeting at a sushi restaurant near campus. When I walked over to pick her up, she gave me a quick tour of her apartment, a small dorm in a block of student flats. She was pursuing an art degree and had a few intricate sketches hung in small frames around the room. It looked like she was musical, too, since there was a small keyboard as well as a ukulele. She mentioned the instruments, saying she loved to play them, as they let her express herself.

'That's so cool,' I said. 'You'll have to teach me! We could do a duet.'

'What? No. I only play with my dad. Music's our thing.'

'Aww, that's so sweet. He comes to visit you?'

'He used to. Then he stopped.'

'Oh, that's a shame – why?'

'He died.'

After an excruciating fifteen seconds or so, I broke the silence to suggest we get a move on for dinner. It was nice out, so we sat in the outdoor portion of the restaurant under a miniature pagoda.

The meal itself went smoothly – no wasabi in the eye, no slippery sashimi, etc. I even felt invincible enough to wheel out old faithful: a full-throated rendition of I Am the Walrus, replete with chopstick canines dangling from under my upper lip. (Not really, although I have no doubt this would have been well received.)

After we finished eating, the waitress offered us a teacupful of fish food to sprinkle into the large, open-topped fish tank along one of the walls behind us. I took a pinch and dropped it in, whereupon one of the bigger koi lunged for it, breaching an inch or two above the surface. Jasmine, who'd stayed behind me at the table, asked me to do it again. I obliged. She then repeated her request, more sternly this time.

'Again. This time, make him splash like before.'

I turned around to see that she was recording all this with her phone.

'Sorry – are you filming me?'

'My friends want to see what you look like. Come on, feed the fish! Don't make it weird.'

I fed the fish in as not-weird a way as possible. As I walked her home, I suggested we stop for a quick cup of tea or coffee at my house. She accepted, a sure sign that we were on the home straight. A small part of me hoped that Fergus's hidden cameras were on so that she could get a taste of her own medicine: *'Oh, by the way, my room-mate likes to watch a live stream of my bedroom. Stop being so weird about it. Besides, you shouldn't flatter yourself. I'm the one he tunes in for.'*

Leslie was the only one home, and he was in his room. I moved a pile of clean laundry off my armchair for her to sit down. Being a gentleman, I would take the straight-backed wooden desk chair. I went out to the kitchen to make the tea, but when I returned to the room, I saw she'd taken my chair instead.

'The armchair is much nicer,' I said. 'You should take it.'

'You left a sock on it. Socks gross me out.'

'Oh, OK, let me get it for you. Here's your tea by the way.'

I passed her the cup. She peered into it and retched.

'This looks gross. Way too dark. I always put more milk in than that.'

'Sure. I'll be back in a second.'

I returned and handed her the hopefully less emetic tea. I was unperturbed. *This was a small price to pay.* I carried myself with a friendly smile throughout, and kept it on as I chuckled and said, 'Jeez, so high maintenance! I'm exhausted.' I flopped theatrically into the armchair.

'What did you just call me?'

'Huh?'

'High maintenance?'

'Oh, yeah, about the tea. I was just joking. Don't worry, I'm fussy about mine, too!'

'You called me high maintenance.'

'Well, yeah, but it was a joke.'

'That's what my ex-boyfriend used to call me.'

'How could I have known that? I really wasn't actually accusing you of being high maintenance – we barely know each other.'

'I think I should leave.'

I walked her home, an awkward fifteen minutes during which I tried to lighten the mood and reassure her just how pleasant the evening had been. With nothing to lose, I leaned in for a kiss when we arrived at her dorm. For whatever reason, she reciprocated – momentarily, before violently pulling away in disgust.

'What's wrong?' I asked.

'Your breath.'

Oh no.

She continued, 'It's ... minty.'

'Oh. Uhh ... Sorry?'

'It's minty, but I didn't see you brush your teeth.'

'I didn't. I grabbed a piece of gum before we left my place.'

'And it was in your mouth while we were kissing?'

'Yeah, I guess so.'

'That's fucking disgusting. Spit it out.'

'Have I offended you?'

'I can't do this anymore.'

'Oh, OK. Sorry you feel that way.'

She spun away and ran inside, never to be heard from again. Looking back, I'm surprised the night lasted as long as it did.

❧

I headed back down into the salt mines (i.e. the dating apps), to find that things were no less maddening here than they had been north of the equator. How on earth could it be that so many of these girls' bio pages were identical – regardless of whether they were in America, England, or South Africa?

- Watch out, I'm fluent in sarcasm.
- If you can't handle me at my worst, you don't deserve me at my best.
- I only matched with you because of your dog.
- Dog mom.
- Mom.

Alternatively: My son is my world. My dog is my world. The baby in the first photo isn't mine, it's my sister's. Work hard, play harder. Laid back. I'm shy, so don't expect me to message first. I'm definitely funnier than you. ENFJ. Don't bother messaging unless you're over six feet tall (or have a beard, or both). Don't bother messaging if you're a Gemini/Taurus/ Scorpio. Touch my butt and buy me pizza. If you like pine- apple on pizza, this isn't going to work. I love pizza. I love food. I love travel. I love my friends. I love to hang out with my friends. My friends made me make this [dating profile]. I don't take this seriously. I only made this account because I was bored. Ready for something real. No more boys. Only real men need apply. I'm only here to make my ex-boyfriend jealous. I'm only here to catch my cheating boyfriend. Empath. INTP. Here for a good time, not a long time. Live fast, die young. You only live once. Just a Bonnie looking for her Clyde. Just peanut butter looking for her jelly. Just a Pam looking for her Jim. If you can't quote every episode of *The Office/ Friends/etc.* don't bother.

All told, it was a bleak state of affairs. In my view, these lines betrayed an appallingly low degree of self-awareness. If you were as clever or funny as you claimed, you'd intuit that this would not be evidenced by picking from the list you saw online of the 'top twenty funniest dating profiles'. After scrolling past the thousandth girl claiming that sarcasm was her second language, I began matching with them for the sole purpose of data collection. Proving my point, but confounding me further, most of them refused to admit they'd plagiarized these lines from anywhere. It was – they swore – pure coincidence. Evidently, these apps impelled all girls to synchronize their mating displays, and adopt these unoriginal, unfunny one-liners, like how girls living in the same house apparently all get their periods at the same time.

Some of the more depressing bios were things like 'I'm actually 19, not 25.' You had to be eighteen to make an account on any of the apps, meaning that loads of these girls had started shopping around when they were several years underage. The ones that really ground my gears were those who offered nothing except 'TIRED OF BEING CHEATED ON', or 'NO MORE ONE-NIGHT STANDS'. Wouldn't a guy who's also looking for something meaningful want more information from his future wife beyond an all-caps admission that she's fed up with getting pumped-and-dumped? I understand that I'm pissing into the wind here. But if I led even one of them to self-reflection, I'll be satisfied.

છે

The gays, meanwhile, seemed to have found a system that worked for them, at least when it came to satisfying their sexual appetites. Compared to the imbalanced dynamics I'd seen in the hetero dating pool (whereby only party was encouraged to exhibit their aggressive horniness and the other had to feign disinterest), it seemed that the objectives here were far more aligned: sexual gratification was the name of the game. According to Leslie, compatibility on this front was in fact so vital that establishing it was the first order of business. The interface for his preferred app looked more like an air-traffic radar, showing a map of the surrounding neighbourhood with pins representing other local users.

On the apps that I'd seen, picture messages were blocked, because guys had been sending out the same picture of their glistening, fully torqued schlong to every girl they matched with before even offering a 'sup'. Here, though, they were encouraged. If the users weren't naked in their avatars, they certainly were in the pictures they sent in lieu of introductory messages. Moreover, there were no restrictions here as to *who* you could message: you didn't need to have matched with each other to chat. For as bold as they seemed online, these men seemed to value privacy in other facets of their lives. Many of them indicated that they were looking for something 'discreet', and would prefer to stay anonymous. Names were sexy innuendo, like 'Dr Big D', or things like 'Hung Bear'. For his sake – Hung Bear's, that is – I should clarify that he isn't claiming to be hung *like* a bear (belying their alpha male reputation, the animals pack surprisingly little heat, proportionally speaking), but rather making reference to 'bear' à la the gay subgroup, of which there were many.

I learned a lot about this taxonomy from Leslie, who had a comprehensive understanding of the 'scene'. You had your aforementioned bears: burly, hairy lumberjack types. Naturally, this meant you had cubs – future bears. Then you had your otters, more toned and svelte than bears but still boasting some (tastefully manicured) body hair. Then there were the twinks, who were smaller still, androgynous, and usually hairless from the eyebrows down. And these just covered the body types. Don't ask me if you can be two types at once: top half bear, bottom half otter, for example. A fascinating thought. This inundation of trivia might feel like a non sequitur. I contend, however, that because so much of this was exhibited/shared with me throughout the year, relaying it to you is my journalistic duty.

Leslie had several repeat callers, none of whom arrived earlier than dusk or left later than dawn. Other than describing their genitalia, Leslie never discussed them further, remaining coy when asked about other facets of their lives. No matter how supportive the rest of us were, he loved reminding us how much he hated living with guys. *Any* guys, but specifically straight ones. Despite railing against the male gaze, toxic masculinity, and so on, he had no qualms ignoring any messages from guys who weren't at least six feet tall or (at least) seven inches 'long'. It seemed like objectification *was* allowed, then, as long as it was intra-sex. Having said that, he encouraged his female friends to be similarly uncompromising when it came to their own suitors, so who knows.

ò

Whatever apps you used, it seemed like this was what amounted to dating these days: a clash of mind games, bluffs, and snap judgements. Romance had, like everything else, been sacrificed at the altar of efficiency: quantified, algorithmed, optimized for maximum convenience. Intangible things like love, trust, and intimacy become transactional. *What return am I going to get for my investment of time, money, and emotion?* We've therefore got no choice but to regard other actors in this 'market' as our opposition. They're trying to get the better of *us* (and us *them*). *Trust no-one.*

The rules of engagement were so pervasive that *not* playing by them was actually seen as off-putting. In multiple instances, I was blocked or ignored by girls after they asked for my Instagram or Snapchat username(s) and I informed them that I didn't have accounts for either, but that I'd be happy to talk on the phone instead. In at least two such cases, I, or my suggestion, was called 'creepy' (things had been going fine until that point, I swear).

It became clear that, despite ostensibly offering empowerment and liberation in the short term, these apps would stiff (as it were) *everyone* in the long run. Despite claiming to value connection and community, they encourage the opposite. If you get bored, annoyed, or if someone better comes along, you don't have to risk confrontation or take accountability, you need only stop responding. The trouble is, there's always going to be a more attractive, interesting option just one swipe or click away. We can now see more attractive naked people in ten minutes than our ancestors could see in ten *lifetimes*. This can't be good for our self-image, nor our expectations. It's almost as if creating emotionally stable adults capable of healthy relationships wasn't really the goal. Compassionate, confident, and healthy people don't make for very reliable consumers, after all.

So, for as much as they disappointed me, I knew that every 'fluent in sarcasm' was a cry for help. I had similar feelings about our enshrinement of long-discontinued shows like *Friends, The Office*, and all the other video games and movie franchises from our childhoods. Did we actually *want* our culture, our very personalities in some cases, to be so dependent on the past? Or was this merely a function of there being no viable alternatives?

There's not just the unsavoury thought that our tastes might be shaped by the market more than we'd like to think. When it comes to this particular seller/consumer relationship, there's an even more predatory element to it. Being raised in an online world where the lines between corporations, and celebrity, and media are so blurred, we take for granted that the producers of culture (especially those from our childhoods) are morally upstanding – or, indeed, that they have any human morals at all. This notion becomes increasingly appealing as our media and advertising grow ever more self-aware, playing on our desire to feel acknowledged and included. Sure enough, we take the bait: we applaud Disney for casting a non-white princess, and we chuckle when the protagonist makes a snarky quip while winking at the camera. We see this as evidence that we are among friends; we are in on the joke.

But I can't blame anybody for sticking to what they know. People have been forced to adopt the same risk aversion as the market itself, and have neither the time nor the disposable income to experiment with anything new or challenging. Nostalgia and irony, then, are just defence mechanisms to help cope with an increasingly hostile world. There's too much bad news to face head-on, so we deflect, clinging to things that have already happened and making silly jokes to disguise the fact that we have nothing to say.

With the end of the academic year upon us, my room-mates and I made our leaving arrangements for the earliest date possible. We were on the home stretch; if we could hold out just a little while longer, we could put this shit-show behind us.

But, as usual, things were never that simple. With only a few weeks to go before I was home and dry, disaster struck: we had another break-in. And this time, it happened while I was home.

(STEAMY) CLIMAX

I'd just returned from a run, taken a quick shower, and was in my room, about to get dressed. Nobody else was home – or so I'd thought. I heard the terrace door, which was right next to mine, slam closed and heard someone stomp in from outside. I froze. Thankfully I'd closed my door out of habit as I'd come into the room.

'Who's in there?' they barked.

I said nothing.

'Who's in there? I can hear you.'

Gruff, masculine, yet somewhat shrill … Wait a second, I knew that voice. Cheryl! But how? Her car hadn't been parked in front of the house or on our street. In any case, she now wrenched my door open and stood in the doorway, her hulking frame blocking it entirely. I managed to snatch my towel off the bed just in time to protect my modesty, if little else. Unfazed, she launched into her interrogation.

'What are you doing here?'

'What do you mean, "What am I doing here?" This is my room! Get out!'

She didn't move, except to fold her arms across her chest.

'In case you didn't know,' she said, 'I own this house, so it's *my* room, technically. These are *all* my rooms. So, tell me.

Why did nobody clean up? I was supposed to give tours here today. I'll have to call them off until tomorrow.'

'But you didn't tell us when you would be here. You have to give us warning.'

'Ha! I'll come and go as I please.'

Glancing down at my feet, she pointed at the trail of droplets that had fallen off me as I'd come in from the bathroom.

'If that water does any damage to the wood, so help me God. Didn't your mother ever show you how to use a towel? Do you need me to show you?'

So this was it. *This* is what the nanny-cams had been set up to capture. This was a page straight out of the Ravesh playbook. *What was it with this family and their fondness for half-naked, post-shower monkey business?* Clearly, she wasn't getting what she needed from her husband (and/or son) and was looking for something new. For obvious reasons, my housemates weren't up to snuff, so she'd turned to this glistening, nubile young himbo (yours truly) in the hopes of resuscitating her dormant love life. I'd never had much of an interest in the whole MILF thing, but if it would soften her up and let us get to the end of the year without further incident, so be it. A dirty job, but somebody had to do it. But what form would our tryst take? A sensual, afternoon-long romp? Or a gruff, no-eye-contact knee-trembler? Maybe she was an exhibitionist like her son. If so, we might find ourselves outside on the terrace, rolling around on the charred turf.

She took a step into my room and opened her mouth to speak. Just then, the front door opened, and Leslie and a friend spilled into the hallway. Cheryl gave me one last up-and-down before spinning on her heels and marching away down the hall.

This visit marked the beginning of the end. The climax, if you will, of our captors' year-long tantric dance. While the lease had said there was no issue with cutting our leases short by a month as long as we provided her with ninety days' notice (which we had), she made her displeasure clear by showing up unannounced to give tours of the apartment on several more occasions. (Whenever I was left unsupervised with prospective renters, I subtly suggested they continue their search elsewhere.) Each time she did this, we reminded her that the lease said she needed to give us forty-eight hours' notice before entering the premises.

'Oh!' she'd say, each time acting more surprised than the last, 'I mentioned to Fergus I'd be stopping by. Did he not pass the news on to the rest of you? He can be such a silly boy.'

ֶ֥

With two weeks to go, we were informed that, before getting our deposits back upon moving out, we'd first be subject to a rigorous, itemized inspection. By this point, my room-mates and I were intimately familiar with every page of our rental agreement. *Nowhere* was such an inspection mentioned. According to her, this didn't matter.

'But you have nothing to compare it to,' we said, 'since you forgot to do a check when we first moved in.'

'Are you arguing with me? You know, your behaviour is typical of tenants who've incurred damages they don't want to be held accountable for. It's almost as if you have something to hide.'

She had Fergus conduct the sweep, of which we were sent the findings few days later. We'd been found guilty of:

- losing a fork and knife (the proof offered for this allegation was 'the cutlery came in a pack of twelve')
- chipping a coffee mug
- breaking a handle on the refrigerator
- seriously damaging the reclining wicker lounge chairs on the terrace outside.

I played along as if any of this was in good faith. I messaged her back to say that we were happy to contribute our share toward the purchase of a new mug and cutlery set. As for the refrigerator handle, it wasn't broken, although some of the plastic had chipped, it was still perfectly functional. And the chairs looked perfectly fine to us. She responded later that evening.

'It's grand of you to be honest about the cutlery. But I'll still be ordering a new handle for the fridge – you three tenants will split the cost. And unless someone confesses to damaging the chairs, you'll be paying to replace them, too.' Attached was the invoice for the handle that was almost three hundred dollars. New chairs would cost over a thousand dollars.

We had no idea where to turn. We called the university but learned they only hosted the rental database and had no connection to any specific landlord who used the site. They referred us to the national association for tenant rights, but the local office had no appointments available for months.

And yet, I was galvanized. I'd finished my last course-work weeks ago, so my every waking second would now be devoted to proving our innocence. I trawled through hundreds of photographs and videos taken by me and anyone I knew who'd visited the house. I researched our fridge's spec-ifications and tracked down a manual online for our specific model. I found the part code for the little plastic strip and tried to find a cheaper replacement online, to no avail. What I did learn was that, miraculously, the brand had an official show-room in downtown Cape Town. I called a taxi immediately. After describing my situation to the sales representative, he told me he'd do his best to source it and get back in touch in due course. He called me that evening to offer the most spec-tacular example of concierge service I'd ever received: 'Good news and bad news,' he said. 'I found the part, but the ship-ping was going to take about six weeks. So I messaged my manager, who was at an appliance trade show in Belgium, where we have our headquarters and main warehouse. As it turns out, he's flying back down to Cape Town tomorrow and can bring the piece with him in his luggage.'

'That's amazing! Thank you.'

He resumed, though his tone became sombre. 'That was the good news. But I'm sorry, sir. Because of how our warranty system works, I have to charge you for the piece.'

Fuck – here it comes, I thought, *and we'd got so painfully close.* He must have heard me wincing as I asked him the damage.

'The charges,' he said, 'will come to … thirty-seven rand. And, again, I'm very sorry.'

This was the equivalent of about two dollars. A miracle. They refused to let me leave some sort of gratuity or pay some share of the shipping charge when I collected the piece at the showroom. I went home and fixed the handle in just a few minutes. Cheryl could barely hide her disappointment as she tested the perfectly restored door during her next visit. Conceding that round, she moved on.

'Well, you're still responsible for the chairs.'

'You mean the chair in the dining room that Fergus tore through with his friends?'

'He and I have already spoken about that. He denies he had anything to do with it. That's none of your concern.'

'What do you mean, "none of my concern"? You've lied to us since the beginning, and you didn't even make him show up for this meeting – how are we supposed to believe you'll hold him accountable?'

'How dare you? I'll have you know he wanted to be here, but he's preparing for a job interview at a law firm. One that helps refugees, by the way.'

'I don't care where he is. *You* picked the time for this meeting. If you'd wanted him to be here, he'd be here. Your precious boy is a sex pest, by the way. He lies around here stroking himself hoping he'll get caught. He even tried to force himself on Leslie. No wonder you have to keep relocating him.'

She'd already been growing red in the face, but the stress had now become too much for her cholesterol-clogged heart. Summoning all her remaining energy, she managed to throw one last haymaker: 'You know what? I used to have no problem with Americans, but I can see now that everything they say about you people is true. Mark my words, I'll *never* work with any of you again.'

After she stormed out, I found myself beaming with pride. Her sentiment, delivered with such theatrical venom, had clearly been brewing for months. I wondered what the exact tipping point had been. While I respected her conviction, I noticed a curious incongruity: despite having lived through apartheid South Africa, this woman had never developed any qualms about taking on the multiple Afrikaner tenants her son had so happily lived with (and doubtless exposed himself to) before. But Americans were where she drew the line? It seemed like the only people she couldn't abide were those who threatened her bottom line.

ॐ

The last time I saw Fergus was a week later. He'd been in hiding, but his mother sent him to collect our keys as we left the apartment for the final time. I couldn't resist: 'Hey man, I was disappointed you missed our meeting with your mom. She really tried to screw us over. We had to fight tooth and nail. I've also been meaning to ask – you have a ton of left-wing political stuff in your room, but don't those guys say that landlords are pretty much the scum of the earth? How do you reconcile that?'

'Yeah, it's difficult. I just try to stay out of the business side of things.'

'But you *didn't* stay out. You sold us down the river.'

'Well, I've helped tenants before when she's done things like this.'

'Really? Who? If *we* didn't seem worth your time, what the hell was she trying to do to the people you *did* help?'

'Well, there was a student from Nigeria. But that was different. You guys had the means to pay for all this stuff. It wouldn't have ruined you.'

'Wait a second. You just go from house to house helping your mom rip students off and, in return, she lets you get away with being a pervert? Do I have that right? You're thirty, and you've now dropped out of three different graduate programmes. How many more years can you guys keep this going?'

'That's not true. I'm only twenty-eight.'

With these encounters out of the way, I could finally exhale. Not a perfect ending, by any means, but it was the best I was going to get.

And maybe we did pay it forward. In the weeks after we'd left, Leslie sent me a screenshot of a dating profile from one of his gay dating apps: a grinning Fergus, standing at attention, decked out in Highland regalia – full kilt and caboodle, as it were. I hoped, for his future room-mates' sakes, that he was finding some release.

THE INNOCENTS ABROAD

I spent my last week in the country with three friends who'd flown down to visit from back home. I'd been desperate to show them this mysterious, far-flung place to which I'd become so attached, and to help them to connect with it as I had. The Garden Route seemed like the best bet; this stretch of coastline had given me more than a lifetime's worth of thrill. Our first stop could only be Shark Island. I promised the guys they had no idea what they were in for. If their dive could be even *half* as exciting as either of mine had, it would be unforgettable. As the captain walked us down toward the water, he saw me pointing out Shark Island to my friends, and chuckled.

'Ha! We aren't going to Shark Island today. Not even close.'

My stomach dropped; the island being just a few minutes' ride away was the only reason I hadn't been violently seasick on the past two trips. I tried to play it cool.

'Oh, interesting. Where's the other island?'

'There'll be no island for us today, but there's a spot a few kilometres away where we've had some luck over the past couple of weeks.'

'Luck? Wasn't it something like an 80 percent success rate?'

'It was. Until the sharks moved.'

361

'Why would they do that? Did the seals all die off or something?'

'It wasn't about food. It was about the orcas. We've started to see them wander into the region over the past couple of months. They're one of the only animals these sharks are scared of, so they left the area. Luckily, we had trackers on a couple of them and found a few who still live further down the coast. We just have to make sure the orcas don't follow us.'

As usual, none of this had been disclosed before we'd paid ('NO REFUNDS!'). I was desperate not to let my friends see my concern; they'd travelled thousands of miles to see me, and in my sales pitch convincing them to visit, their survival had been more or less implied.

Our route to the secret dive location was circuitous and erratic, as if we really *were* trying to lose a tail. I've never been more seasick in my life. It didn't help that the previous days had been stormy, the water was still choppy and distressed. The sky was troubled in its own right – low, dense, heavy, a shade of pewter barely lighter than the water's. This worsened matters by obfuscating the horizon, offering us no relief, no reassurance; pressing us into the water.

For the first portion of the ride, we hugged the coastline, skirting around Shark Island before heading outward to the open sea. The island lingered behind us for a time before gradually disappearing into the fog along with the rest of the mainland. It felt like we spent almost two hours on the rickety vessel, which had looked far sturdier back at the port. I spent most of the time hunched over its side, trying not to vomit. I wasn't even left to suffer in peace: about half the group was stricken by nausea too. While the others formed a line and took turns to expel their continental breakfasts, I refused to look up from my white knuckles and the chrome railing. This gave me more than enough time to reflect on the captain's throwaway comments back on the dock …

The great whites, who'd proven themselves to be perfectly capable of inciting terror, had apparently been dethroned by creatures that could *literally* swim rings around them. Sure, I'd heard the claim that no humans had ever been killed by an orca in the wild, but absence of evidence is *not* evidence of absence; all that meant was that no attack had been *recorded*. Regardless, if our time in TIA-land had taught us nothing else, there's a first time for everything. Humans had already caused countless unforeseen and unprecedented changes to animal migratory patterns and behaviour by destroying habitats and biodiversity. Some animals were already hitting back: for example, elephants in Indonesia raiding farms that had encroached on their territory. If orcas were half as clever as our research suggested, they'd have long ago figured out that we represented their certain extinction/enslavement, and had decided to strike back. We had a track record of destabilizing regions on land and inciting radicalized militia groups. Why would things be any different here at sea? Could we really rule out that fundamentalist splinter cells were now roaming the oceans, recruiting sympathisers and killing dissidents along the way? I'm not sure we could. In my estimation, there was a decent chance that any remaining shark within a few hundred nautical miles was now a card-carrying insurgent, highly trained in guerrilla warfare and 'enhanced interrogation'. These were not the same animals who'd clumsily wedged themselves into my cage in the past. That old guard had been docile, compliant, all but domesticated by comparison; those bootlickers had been the first to go. Thinking back to the scene of the baby seal's lurid mutilation, everything now made sense: that was the sharks' version of a Colombian necktie.

As I glanced back toward Shark Island, my stomach dropped. Maybe it was a nausea-induced hallucination, but I was sure I saw a creature, too large to be a seagull, flying through the air. I could vividly picture the scene: another young seal – this one detained after being caught slurping some 'man-chum' – being flung back and forth above the frigid water by two neophyte sharks, their sinewy muscle rippling beneath abrasive, metallic skin. The pup yelps pleadingly as he soars between his tormentors, trapped in a doomed coin flip: heads they win, tails you lose. He watches on as his abdomen becomes a colander, his innards spilling from his plump torso in thick ribbons like minced beef falling from a grinder. I now understood why the captain had been so coy about our dive location – any loose lips and we'd get the same treatment.

We drifted to a stop and dropped anchor in the open ocean, the nearest land a vague sliver on the horizon. This gave the feeling we were more vulnerable than we'd been on my previous dives, when we'd been much closer to shore, or at least Shark Island. It's not as if we could have reached either of those in the event of emergency, but the proximity had provided some sort of reassurance, placebo or not. At any rate, I was so nauseated at this point that getting into the freezing, murky water was a relief.

Normally, in the five- or ten-minute turn in the cage, the sharks approached the boat once or twice a minute – a hit-rate I had assured my friends of. Alas, it seemed the orcas' scare tactics had been successful. We waited as long as we could, but only one measly shark risked associating with us, this one much thinner and shorter than the ones I'd seen before. She circled the boat a few times, warily inching closer, before stalling about five metres away. She lingered just long enough to have a taste of chum before slinking off into the dark. Anticlimactic as the dive was, it was far preferable to spending another second on the boat. Sadly, there was no option but to suffer for another hour and a half on board, with what felt like acute hypothermia now kicking in for good measure.

᠎ஃ

The next morning, we had a half-day safari planned, which I was hoping would restore morale. While we were en route, however, the game reserve called to tell me there had been an error with my reservation: our tickets been cancelled and all seats had been filled. Our only other option, the secretary said, was the sunset safari, so I signed us up. I didn't know what to expect, as I hadn't seen information for any such expedition on their website. I didn't know it yet, but a bureaucratic error had finally worked in my favour.

We got to the reserve the next afternoon and were introduced to the guide, whose name was Innocent. Names like his were common, sometimes they were direct translations from languages such as Zulu or Xhosa, and were very often nouns, adjectives, or even phrases. In the weeks leading up to this safari, I'd met a Godknows, an Admire, a Shepherd, and a Happymore, along with a few Bondos and Phaphamas. I loved these unconventional names, compared to the sterile offerings of suburban America (case in point: our group included a Jack, a Dan, an Al, and a Jeff).

As we climbed into our seats, Innocent loaded a large crate into the storage boot at the back of the vehicle. 'Can't forget the snacks!'

'We're going to feed the animals?'

He laughed. 'No, no. They're for you!'

Two other tour groups had dropped out, which meant my friends and I shared the vehicle with just a middle-aged couple and their young son who sat in the back and kept to themselves. Toward the front of the truck, we kept up a running conversation with Innocent as we drove off into the reserve. Not only was he immensely knowledgeable about the reserve's wildlife on a biological level, he was also familiar with their individual behaviour. On some days, he told us, certain giraffes or rhinos were much harder to find, or even avoided the humans altogether. That afternoon, however, we'd be able to see even the more reclusive characters as we were in a smaller truck and could more easily traverse the treacherous hills and take some of the unofficial trails through the park.

After touring for about two hours, Innocent swung us around a hairpin turn and we climbed out of a narrow gorge, where we'd been watching a family of elephants graze. We arrived on the highest point of the reserve, a plateau with tussocks of waist-high grasses that fed small herds of spring-bok and zebra. In the distance, spanning the northern horizon and rising much farther into the sky than the foothills, were the Outeniqua Mountains. He parked the truck and walked around to open the boot, passing out thick fleece blankets for us to unroll on the ground, and some smaller ones to drape over ourselves; it was crisp in the late afternoon, especially up on the plain, and the temperature was dropping as the sun got lower. He popped another large compartment in the boot to reveal a minibar and told us to help ourselves.

We enjoyed several rounds of drinks while taking in our surroundings, the animals grazing peacefully and regarding us with curiosity. To mingle with the wildlife on this high, remote plain was thrilling. We weren't quite in the bush, of course, but this was as close as we could get. The main lodge was a few kilometres away and completely out of sight, and there was no fencing, roads, nor any other evidence of humans but for a lone power line suspended above the wheat-gold plateau. The wire, a silvery thread against the forbidding backdrop of the mountains, ran for a kilometre or so before dipping out of sight in the northeast.

As it got darker, I asked Innocent the whereabouts of the reserve's lions. I tried to sound casual, not bringing up what I'd learned a few years earlier about their night vision (or, for that matter, their mating habits). Thankfully, we were safe: the lions were kept in a separate segment of the park.

Eventually, it was time to make our way back to the lodge. Loathe to waste an open bar, I suggested to the guys that we mix some drinks for the road. In a rush, I poured us each several glugs of Amarula, a delicious South African liqueur made from the fruit of the marula tree (which, legend has it, are also a favourite of the region's elephants). Whether due to the dark or my buzz, I started pouring before realizing Al and I still had about half of our gin and tonics remaining in our glasses. When we clambered into the truck, the family now sat in the front, so we foolishly chose the back row. We tasted our concoctions, expecting the worst, but were pleasantly surprised; it seemed the ingredients had a tenuous equilibrium, rendered into a kind of smoothie by the gentle vibration of the all-terrain vehicle. But this lasted for only a few seconds. As we trundled along, Al looked down into his brass goblet.

'Oh, that's weird. What happened to our drinks? It's like they've curdled or something.'

He was right: the gravelly rumbling had agitated our cocktails to disastrous effect; thick globules now swam in a beige, milky bile. But the taste was still fine; we had to see it through. I looked up ahead. The headlights revealed that there were only a few metres left of relatively flat grassland – the treacherous ravines were yet to come.

Sure enough, we violently pitched downward seconds later to access what looked like a goat path snaking down the slope. Innocent, evidently a Formula One enthusiast, seemed to relish taking this challenging circuit in the dark. Now that we had no sightseeing to accommodate, we went much faster than we had on the outbound leg. Without seatbelts to restrain us, centrifugal force slid us across the cheap, slippery vinyl and smashed us into each other as we hurtled around hairpin turns. The back row, as we ought to have learned from countless school bus rides, is hypersensitive; we were bucked into the air by every pothole and bump. Before long, our shirts and laps were drenched from the spillage, but we'd stopped caring and were in hysterical laughter as we succumbed theatrically to the truck's every whim. We hurtled toward the pit of one valley, where a narrow stream wound through beach-ball-sized boulders and skirted around a dense grove of low vegetation. Innocent lurched us to a halt by a wobbly wooden footbridge.

'If we're very lucky,' he whispered, 'we'll see one of our new arrivals.'

The light had faded, and the rocks, trees, and undergrowth had all sunk into shades of murky brown. Unless the new arrival was wearing a lit headlamp, the odds of us being able to make it out were slim. But from somewhere behind the bushes, a miniature rhinoceros trotted out confidently, followed closely behind by an adult. This was lovely to see, especially after the heart-rending backstory Chet and I had heard on my first game drive. I knew better than to ask Innocent anything about the rhinos, including how many now inhabited the park. The man in the front row had apparently missed this memo, and I was pleased to see Innocent adopt the same coyness as my first safari guide had:

'How many rhinos? Well, as you can see, there's two.' He gave the group a cheeky grin before turning back toward the animals. We eventually got back to the lodge and walked through the main hall, which triggered another wave of cackling as we imagined how we must have looked: two of us had spent the drive clinging to each other and vomiting into each other's laps, while the other two had apparently been perfectly content to remain in the splash zone. We drove back to Cape Town, buoyed by our expedition into the wild.

๛

The safari drives I had done before, albeit perfectly enjoyable, had lacked this crucial element. Not the coagulated slop, that is, but the intimacy with nature, however contrived it may have been. Private, exclusive experiences *are* entirely possible, of course, but prohibitive. You could, for example, hire a lone guide to take you into the heart of Kruger National Park to track a cheetah, but this would be horribly costly in terms of time and money.

Back in the city the next day, my feelings were reaffirmed as we hiked Lion's Head. This climb is a critical item on any visitor's agenda, and understandably so: the mountain, with its panoramic views, is perfectly situated, access to the trail is free, and it's a relatively easy climb. Good weather sees hundreds or even thousands of hikers summit and descend on any given day, navigating each other as much as the terrain. Any large branches and roots growing near the single-lane path are used as handholds and footholds, rendering the wood glossy and slippery, as if sandpapered and varnished. It's especially busy around sunset.

At dusk, the shadow of Lion's Head is projected down into the City Bowl, like the cartoonish shadow of a villain towering behind an oblivious child. The houses in this unlucky triangle are benighted long before the rest of the city – which must be aggravating with electricity so scarce. Soon, though, the sun drops lower, and their neighbours join them in darkness. Thin lines of fluorescent red and white crawl slowly over the pass to and from Cape Town's famous, palm tree-lined, sunset-facing beach, Camps Bay, which milks every second of natural light before illuminating itself like the rest of the city. (But you don't want to stay too long, as muggings are common after dark.)

The pinched summit of Lion's Head, rising to almost 700 metres above sea level, is high enough to distance you from civilization, in theory. Any humans down below are barely visible, and sound from the streets is muffled by the wind and the static of the Atlantic waves that pummel the surrounding coastline.

But you're soon reminded that you can't outrun the modern world. As sunset neared, I began to notice a frenzied buzzing loud enough to carry over the wind, consistent in pitch but increasing in volume as more hikers gathered at the summit. This was no swarm of bees or locusts, however, but rather a dozen or so electric drones, controlled by their owners via phones or remote controls. Some hovered directly above, uncannily still, like birds of prey riding a thermal, while others had been sent out on various orbits above us.

Incidentally, if we had ventured one peak higher to Table Mountain, we would have avoided all this, as drones are banned. Not that it doesn't have its own share of human sprawl to deal with: a concrete lodge on the nearest corner of the plateau runs thick steel cables to an even larger building at the base of the slope, carrying gondolas of people to the summit and back. Operations continue until an hour after dark to accommodate not only the sunset photographers, but the dinner rush. A few hundred square metres have been paved atop the mountain to provide a foundation for gift shops, bathrooms, and a large dining hall. Joining the tourists for mealtime are the aforementioned dassies who, in ones and twos, tentatively approach anyone sitting at the edges hoping they've got a few extra potato chips or pieces of trail mix to spare (and they are usually rewarded).

A rebuttal to my grouchiness would probably sound something like: 'We ought to open our minds to celebrate such a beautiful sunset or trail with even *more* beautiful strangers.' I'll grant that most of the Cape and the Garden Route's offerings still manage to be awe-inspiring and enjoyable even amid throngs of tourists, and that most inconveniences faced at these sites are trivial, such as waiting for people to get out of your way or to stop blocking your view. Still, I found myself wary of this 'the more the merrier' angle, as it feels too close for comfort to the logic of the guy who brings his guitar to a party uninvited. *'What, you don't like music? Who doesn't like music? Don't be a party pooper!'* – you give an inch, they'll take a mile. And now you're stuck listening to Wonderwall for the fifth time in a row.

Anyway, while the window to rattle off petty grievances remains ajar, I also believe that the able-bodied ought to climb Table Mountain, even just *once*, before they're allowed to take the gondola up. Along with terrific views along the way, its plenitude of trails dilute the crowds and offer a more rewarding experience, and the easiest of them only takes about an hour. *Come on, you can do it!*

Believe it or not, my gripe isn't with tourists in particular (although I can't deny hoping those drones would kamikaze their owners). As someone who wants these same experiences for myself, I'm a hypocrite. If it were up to me, I'd have unfettered, private access to any of these parks. But perhaps this was a moot point: I'd been reminded on countless occasions to be wary of secluded areas, even those near the city. At the entrance of every hike or park, staff or other visitors made sure to notify me there had been several instances of disturbing crime in the nearby area recently, or that muggers liked to target these areas given the lack of potential witnesses. *Safety in numbers.*

If total tranquillity and solitude in nature is an unrealistic expectation, perhaps we might compromise and consider natural wonders sacred by default. We ought to have the choice to opt out of being filmed by a fleet of drones, captured in dozens of selfies and videos, and soundtracked by music blaring from a speaker being lugged up the mountain by the dude a few metres ahead of us on a trail. Admittedly, the man running up Lion's Head in an inflatable dinosaur costume was a hilarious sight. So, too, on a different occasion, was the bachelor party in full-body gorilla suits. But, still…

Anyway, for just a few hours on that little safari drive, we got to experience what life would be like in my Goldilocks world.

❧

On my friends' last day, we went for a hike along the ridge above Camps Bay. Afterward, we decided to stop for a drink at a hillside pub across the road from the trailhead. Before the waitress arrived to take our orders, I sought out the bathroom, which was just beyond the bar's outdoor seating area that overlooked the City Bowl. A gay couple was sitting at one of the tables, and as I walked past I instantly recognized a distinct, nasally Australian accent: it was Teddy's, the volunteer for whose date I'd helped rehearse. (I didn't bring this up in front of his current partner.) He hadn't been back here since that trip, he said, and was just visiting for a week or so before going home.

Impelled by nostalgia, I returned to Muizenberg after my friends left for the first time since my original stint in the country five years earlier. I walked along the town's beach promenade where, one by one, I took in all the familiar places.

The laundry owned by the two old, coloured ladies where I'd brought my dirty clothes, balancing a full contractor bag between the handles of my bike on the wobbly ride across town. As useful as the bike had been to shuttle me safely around the neighbourhood, it had no shock absorption, and the jarring of its hard plastic handlebars gave me deep bone bruises in the heels of my hands which remained sore months after I'd left. Before leaving, I ended up giving the bike back to the girl downstairs; I wondered if she'd suffered, too.

The surf shop where I'd paid for my first – and last – surfboard rental. I'd been inspired by a small class of wet-suited black kids idling offshore. They looked relaxed and confident as they sat astride their small boards, legs dangling into the water. When they sensed the next wave was going to be suitable, they quickly lay forward to paddle into position. As soon as they caught the lip, they sprung to their feet and sliced artfully along the inner face of the wave as it broke behind them and carried them to shore. Some used the slope as a ramp, turning upward and letting their momentum shoot them airborne, performing a spin or flip before landing gracefully. I watched for a few minutes, mesmerized, before resolving to try it myself. The rental agent had offered a discount on a lesson with an instructor, but I figured the kids hadn't done anything I couldn't teach myself by trial and error (or just brute force) in the half-hour time slot. I was wrong, *very* wrong – I barely even made it to the water. I couldn't corral the special beginner's board (in my defence, it was about twice my height), which repeatedly ploughed me over into the sand as it caught the offshore breeze. The man had told me not to bother with smaller waves, as they lacked sufficient momentum and would be more difficult to catch as a newbie. I paddled out and tried a couple of times but soon gave up, opting to just use my cumbersome slab as a boogie board. The massive waves still had their way with me, putting me through pummelling spin cycles on their way into shore. I quit after twenty minutes.

I also saw the Indian restaurant where the Soul Sisters had congregated on their charismatic leader's last evening. And the food market one street farther back where the volunteers had gathered every Friday evening to try samples of South African and international dishes.

Across the road, running perpendicular to the boardwalk and south along the peninsula, were the train tracks that had carried me to the tennis club.

Finally, the internet café where I'd watched that tennis match, although what I really remembered from that afternoon was the domestic drama at that table next to mine. I peered inside.

No ... it couldn't be ...

BACK TO SQUARE ONE

The very same father and son, sitting at the exact table they'd shared years before. The dad was wearing the same spectacles and looked healthy, although his once-brown hair had fully given over to the flecks of grey I'd noticed when I first saw him. The son had grown up, but still looked to be a year or two away from needing to shave. From outside on the pavement, they seemed to share an affectionate rapport; the son was animatedly telling a story between spoonfuls of ice cream, his dad hanging on to every word.

Narrative-wise, these two made my job pretty easy: forgiveness, reconciliation, salvaging a positive outcome from unlikely circumstances. And what about their country? Ideally, there was a nice little bow I could put on that side of things, too.

Alas, in this regard the last few years didn't appear to have been nearly as formative. Power cuts were increasingly frequent, crime and HIV/AIDS rates were still high, the townships didn't appear to have shrunk. Any shift, if it had occurred, had been on my end, in my perception of it all: while it was all no less depressing at the objective level, I'd certainly become desensitized. Perhaps this, at scale, is what happens over the years: the unconscionable gradually becomes conscionable.

Oh, well. It was time to go back to the real world. What would it have in store? I guess I was half-hoping that I'd get a hero's welcome. Doors and windows of opportunity would open for me, unbidden; I'd be flooded with offers for gainful, fulfilling employment, and fanned with palm fronds by beautiful women.

Clearly I'd been out of the loop: the consensus was that a masters degree was the new bachelors degree, barely qualifying you for even entry-level work. Yet again, I found myself back in farm country, helping my mom unload groceries and running errands for my dad. I was learning the same lessons I had every other time: you're not as unique and interesting as you think you are, and nothing really changes while you're away. People didn't even know I'd left. When I told them where I'd been, they still looked as bewildered as they had before my first trip. Africa was a place of mud huts and warlords and man-eating lions. Once we got past that, the conversation was steered back to familiar waters: *whether* I was married and had children, *why* I wasn't yet married with children, and *when* was I planning to get married and have children. Obviously, I had no more clarity on this topic than I'd had *before* I'd dropped out of university. Less, if anything. It seemed like I wasn't alone birth-rates were dropping in developed countries around the world. Even ambitions as modest as settling down weren't as future-proof as advertised.

Even the Amish, whose raison d'être was to resist the temptations of the outside world, had been compromised.

More and more of their historic farms and homesteads were being bought up by investment firms, to be razed and levelled for strip malls, gas stations, and drive-thru fast-food restaurants. The rot had set in at the micro-level, too. There was a famous farmers' market downtown where the Amish would flog their fruit, veg, and their signature patchwork quilts. Growing up, most of the painted wooden signs above their stalls had said things like 'Eli Stoltzfus & Sons, Organic Produce', or 'Jebediah Smucker, Meat and Dairy'. Now, though, I noticed that many of them had been replaced by cheap plastic tarps printed with things like: 'Quaint-Amish-Souvenirs.com'. And their merchandise was drifting into dangerously heretical territory – many of the small towns around Lancaster County were ripe for double entendre, so maybe it was only a matter of time. You started to see cheap T-shirts and fridge magnets saying things like: *I started off in Virginville, got stuck in Blue Ball, but then I took my wife to Intercourse ... and now I'm in Paradise!*

I'm not sure which thought was worse: that the Amish had become self-aware and made all this themselves, or that truckloads of sweatshop-made junk had been foisted upon them by some cynical outside investor. Either way, it didn't bode well. If *they* had sold out, the rest of us didn't stand a chance.

And things soon took a turn for the worse. I'd barely been home a couple of months before we got a global pandemic, during which the world's richest few percent managed to become significantly more so, at the expense of the remaining seven billion. The healthcare systems in both the US and the

UK were brought to their knees, with politicians capitalizing on fear and confusion for their own political and financial gain. Wages and supply chains froze, while asset prices and inflation went up, leading to cost of living, housing, and food crises. In the US, racial tensions bubbled over after another cop killed another unarmed black man. We then got an election between two men who would probably have failed a driver's test, following which the loser called into question the integrity of the result and incited his supporters to break into the Capitol.

After a couple of years, things gradually returned to normal. But what did 'normal' actually mean? We were the Top Dog, so 'business as usual' was inherently a good thing, right? All the not-so-good sounding stuff was just the price of freedom. But that thought from South Africa – about the unconscionable gradually becoming conscionable – had stuck with me. Was that a phenomenon unique to the 'third world'? Or, with the help of the pandemic, had we been forced to accept just how precarious our own circumstances were? How dependent were we on these people and places we knew nothing about?

The soul-sucking online dating 'market', the mega-budget superhero spin-offs, the Himalayan goats, the guilt-ridden voluntourists, the predatory landlady and son, the geriatric politicians, Zimmer's app (and all the others like it), the needless X-rays, the hornless (and horn-y) rhinos, the townships, the Amish souvenirs, the having to beg my co-workers for my eighty-seven cents, the radicalized orcas, the mutilated seals – *all of it*. Could it really be a coincidence that everything had become a race to the bottom, a tragedy of the commons, or some combination of both?

There's simply no such thing as 'someone else's problem' anymore; the challenges we face are too interwoven, too complex.

What this means, then, is that the 'third world' conditions we see in places like South Africa are not our past, but a glimpse into our *future*. This is what happens when regulation is scrapped and the relentless pursuit of narrow interests is left unchecked. Invariably, things tend in the same direction: a near-feudal power structure in which not only politics, media, and culture, but things like food, water, shelter, and law enforcement are consolidated into the hands of fewer and fewer private actors. Why would we be immune? We are beholden to the same hierarchies and incentive structures, after all.

We may think that the last great land grab was for Africa, with fat, jowly Belgians, French, and English sitting around an unfurled map in some smoky boardroom, but a similar process is happening now, with us as the subject. We're like whales to a tribe of Inuit – every last scrap of our bodies, minds, and time is fair game. But even this analogy isn't perfect; those peoples still had reverence for their felled prey. The contempt with which *we're* regarded is reflected everywhere: in our architecture, art, and media. *Give 'em cheap, disposable, and easily replaceable; they won't mind.* And we *don't* seem to mind.

Sure enough, it seemed like the warnings of the pandemic would go unheeded: a return to a mask-free world was enough to placate the masses. Despite now being a convicted felon, Trump managed to win another general election, capitalising on the media's willingness to platform him and legitimize him yet again, and the democrats' insistence on sticking with Joe Biden until it was too late. England, meanwhile, cycled through several corporate-backed prime ministers of its own. With millions of their own citizens living one medical emergency away from bankruptcy, both governments nevertheless spent billions of dollars defending their interests abroad, funding the Ukraine in a proxy war against Vladimir Putin, while helping Israel carry out what all international human rights advocacy groups deemed to be a genocide.

Maybe things would have to get a whole lot worse before they got better.

❧

There's a lot to chew on here. How can we hope to function without getting too bogged down? The first step, I think, is to stop the bleeding. To do this, we can take a page or two out of the Amish-country playbook. They seemed to have no shortage of things like optimism and community. Maybe there was something to the whole 'wilful ignorance' thing; it sure would be nice to not live in a constant state of overstimulation and fear of missing out. I'm not saying we need to go too hardcore, ditching our cars, light bulbs and any clothing with zippers, only that spending our waking moments reading through the world's most distressing cultural and geopolitical issues may not be as helpful as we think. 'Maximally informed' and 'wise' might not be as synonymous as they appear; the same goes for 'consuming content' and 'activism'. Again, it's not that our frustration and anguish are irrational – anything but. It's just that, as individuals, these are impossible millstones to carry if we're to have any hope of preserving our health and sanity.

Turning down the volume of the outside world is a good start. As far as I could see, though, the real gold dust was something else: their belief in something bigger. The thing is, we've debunked all the fairy tales, we've got all the facts and figures before us, we've optimized and streamlined everything, yet we remain anxious, lonely, and dissatisfied. Turns out, it's a big, lonely world to navigate without any narrative undergirding.

EVERYTHING HAPPENS
FOR A REASON
(AND OTHER CONSOLATIONS)

And this puts us in quite the dilemma. For a hyper-individualistic, anti-authoritarian, highly-rational society, the thought of ceding control of our lives to a force we don't fully understand is kind of a deal-breaker. We may not be happy with our lives, but we wouldn't trade being the main characters for anything. Self-mythology is a hell of a drug, and we indulge ourselves constantly, constructing elaborate backstories to conceal our laziness, insecurities, and our shameful ulterior motives. We're especially skilled with the benefit of hindsight, whereupon we can sift back through our memories and retrofit whatever scaffolding we want.

Over time, our stories acquire a momentum of their own, an inevitability, and it's impossible to imagine things having turned out any other way; we gloss over or forget just how unstable their foundations really are. In my case, 'three stints, a dash of philanthropy, and a couple of degrees', sounds solid enough. But if that first flight to Cape Town had entailed a prohibitive eighteen-hour layover in Ethiopia or Doha, say, or if the only volunteering role available had been rebuilding a community centre somewhere in the slums of Lima, my 'passion for saving the children' might well have been shelved indefinitely. (That stays between us, I hope.) Oh – but if I *had* gone to Lima, you'd better believe I'd be telling you about my aunt who climbed Machu Picchu, and about my first Tintin book, *Prisoners of the Sun*, in which he and Captain Haddock get captured by a lost tribe of Incas deep in the Andes (I'd read this one until the pages literally fell out). But I haven't been to Peru – at least not yet – so Tintin and all the other potential contributors for *that* story remain dormant, gathering dust until they're called into action.

❧

This is something I'd been thinking about a lot since coming home, as I struggled to figure out what the last few years had meant. There had to be *something* I could take from them (other than a superiority complex and an apparently worthless degree). I couldn't just have been using Africa to avoid confronting the paralyzing and growing suspicion that I would never be a functioning, independent, 'productive' adult. *But what about that last day in Muizenberg?* Seeing the father and son *had* to be some sort of sign, didn't it? Some cryptic message from the universe confirming how special and unique I was?

Maybe not; maybe it was 'just chance'. Fine. Even then, though, did that really do the story (or me) justice? I hadn't just been flotsam, floating aimlessly along the river of fate. Had I not at least put my thumb on the scale? And what about the father and son? While I'd left to carry on with my haphazard navigation of the world, life in their sleepy suburb had gone on and they, too, had lived through the last half-decade. What trials had *they* been through? What choices had *they* made? They'd played a role in this, too; was that not worth celebrating? Say the dad had only picked the café based on nothing but a quick web search for the best ice cream in town. Whether or not his decision's rationale was trivial, its consequences – years of shared memories and experiences – were anything but. Would it be so narcissistic, then, if he credited himself, at least in part, with the outcome? I sure hope not. The thing is, I fear this sort of revisionism is inevitable, given the hand we're dealt, as simple creatures tasked with making sense of an impossibly chaotic and hostile world.

If our love of narrative is baked in, it's on us to figure out how to channel it for good. Not only is it worth a shot, but, as we career headlong toward climate catastrophe against ever-lengthening odds, there's no other choice; we'll need to work with the tools we've got. This means accepting that our capacity – indeed, our eagerness – to suspend rational thought is not a relic of our benighted past, but an invaluable tool, and one that is already being used against us.

Indeed, it can't be overstated that the power structures around us are as much *belief* systems as anything else. Though ostensibly secular and highly-rational, they are of course inherently contradictory, and therefore require a theological framework: they rely on mythology, on superstition, on faith, on the way they make people *feel* (these feelings might not even be pleasant, but they're familiar). As such, attempting to combat these power structures at the intellectual level alone is futile. There must be something more; a north star, a unifying message, a positive future to work toward.

Which brings us back to the Amish point: the gold dust isn't really the specific beliefs themselves, but merely their willingness to hold them; to believe in something. The willingness to set aside short term, individual desires in service of a shared objective on behalf of people they might never meet. This is what has been taken from us: the ability to even *conceive* of a better world, let alone the motivation to work toward it.

So, in the fight to stave off nihilism and despair, our challenge will be to have faith. Faith that, despite all evidence to the contrary, this is not the natural, inevitable state of things. Faith that, as individuals, our decisions do matter; that our contributions can and will accumulate.

This is what we'll have to lean on: since we don't know how our actions, however seemingly inconsequential, will impact our lives and others for the better, we have no excuse but to play our part. In a world of perfectionism, decision paralysis, and risk-aversion, our challenge is to take a step, however ungainly, into the unknown, even if – no, *especially* if – we don't know what the last step will look like. And I'm not even saying you can't be weird, neurotic, and grumpy along the way. If anything, it's *more* important for us, the habitual ruminators and armchair-theorizers, to stop our mindless scrolling and go explore.

Obtaining a wider breadth of experience isn't just résumé-filler; it allows us more creative license as we make sense of the world and relay what we've learned to others. And this is a vital skillset, as the quality of our interpersonal relationships plays a primary role in determining our happiness. Contrary to what we're taught, we are not 'mercenaries', 'lone wolves', or singular, isolated nodes. We are deeply sensitive and social creatures with very serious – and heartbreakingly basic – needs, such as reassurance, laughter, touch. We must therefore ignore any attempts to dissuade us from meeting these needs on the basis that they are childish, self-indulgent, or a non-optimal use of time. We must remember that our disenchantment, anxiety, and low-level sense of dread are not our fault, nor are the shame and alienation we feel for experiencing these symptoms. The fact that we do not have the tools to address these things in any meaningful way is no accident. This is what must be rebuilt: a shared understanding; an awareness. I guess the takeaway here is that, if we're all going to be unreliable narrators, let us at least be inspiring, for God's sake.

As long as we remain gracious, of course. Whether we like it or not, our lives are shaped by vectors far beyond the scope of our existence. And we also know that it can be the most fleeting moments and serendipities that have the most profound impact. As far as we know, then, every link in the chain of successive events is of equal importance.

It's for this reason (and I'm sure he'll be relieved to hear this), that I've come full circle, and decided to forgive my German nemesis. We simply can't rule out that, *without* Herr Zimmer (and Mr Simon) my time in South Africa – and, just as crucially, my years of being the cool guy who kept going to South Africa – might never have happened. No hikes, then. No Haka, no sharks, no skinhead baboons, no Pastor Samuels, no Innocent, no caveman molesting himself athwart the sex-bed. Unthinkable! If nothing else, that's material I can't afford to lose.

ACKNOWLEDGEMENTS

A massive thank you to everyone who helped with the creation and production of this book; research, writing, revisions, read-throughs, and everything else: Robin, Ron, Sara L., Richard R., Ryan Q., Daniel M., Eric A., Jon A. Shathley, Liv, Claudia, Mr. Flea, Oli, Sarah B. (and the rest of the Lesvos Class of '23).

And to anyone who isn't referenced in the book but was a part of these adventures: Stevie H., Alasdair F., Kirsten T., Sophie, and Val.

To my parents, of course. And a mention for Cora G.

But the biggest hug of all goes to M. Bao.

AUTHOR BIO

Jack grew up in Lancaster, Pennsylvania. Now based in London, he is hard at work on his next projects. When not writing, you'll find him re-watching old Roger Federer matches (and reminding himself to not cry because it's over, but to smile because it happened).